Electron Microscopy of Materials

An Introduction

MATERIALS SCIENCE AND TECHNOLOGY

EDITOR

A. S. NOWICK
Henry Krumb School of Mines
Columbia University
New York, New York

A. S. Nowick and B. S. Berry, ANELASTIC RELAXATION IN CRYSTALLINE SOLIDS, 1972

E. A. Nesbitt and J. H. Wernick, RARE EARTH PERMANENT MAGNETS, 1973

W. E. Wallace, RARE EARTH INTERMETALLICS, 1973

J. C. Phillips, BONDS AND BANDS IN SEMICONDUCTORS, 1973

J. H. Richardson and R. V. Peterson (editors), SYSTEMATIC MATERIALS ANALYSIS, VOLUMES I, II, AND III, 1974; IV, 1978

A. J. Freeman and J. B. Darby, Jr. (editors), THE ACTINIDES: ELECTRONIC STRUCTURE AND RELATED PROPERTIES, VOLUMES I AND II, 1974

A. S. Nowick and J. J. Burton (editors), DIFFUSION IN SOLIDS: RECENT DEVELOPMENTS, 1975

J. W. Matthews (editor), EPITAXIAL GROWTH, PARTS A AND B, 1975

J. M. Blakely (editor), SURFACE PHYSICS OF MATERIALS, VOLUMES I AND II, 1975

G. A. Chadwick and D. A. Smith (editors), GRAIN BOUNDARY STRUCTURE AND PROPERTIES, 1975

John W. Hastie, HIGH TEMPERATURE VAPORS: SCIENCE AND TECHNOLOGY, 1975

John K. Tien and George S. Ansell (editors), ALLOY AND MICROSTRUCTURAL DESIGN, 1976

M. T. Sprackling, THE PLASTIC DEFORMATION OF SIMPLE IONIC CRYSTALS, 1976

James J. Burton and Robert L. Garten (editors), ADVANCED MATERIALS IN CATALYSIS, 1977

Gerald Burns, INTRODUCTION TO GROUP THEORY WITH APPLICATIONS, 1977

L. H. Schwartz and J. B. Cohen, DIFFRACTION FROM MATERIALS, 1977

Zenji Nishiyama, MARTENSITIC TRANSFORMATION, 1978

Paul Hagenmuller and W. van Gool (editors), SOLID ELECTROLYTES: GENERAL PRINCIPLES, CHARACTERIZATION, MATERIALS, APPLICATIONS, 1978

G. G. Libowitz and M. S. Whittingham, MATERIALS SCIENCE IN ENERGY TECHNOLOGY, 1978

Otto Buck, John K. Tien, and Harris L. Marcus (editors), ELECTRON AND POSITRON SPECTROSCOPIES IN MATERIALS SCIENCE AND ENGINEERING, 1979

Lawrence L. Kazmerski (editor), POLYCRYSTALLINE AND AMORPHOUS THIN FILMS AND DEVICES, 1980

Manfred von Heimendahl, ELECTRON MICROSCOPY OF MATERIALS: AN INTRODUCTION, 1980

Electron Microscopy of Materials

An Introduction

MANFRED VON HEIMENDAHL

Institut für Werkstoffwissenshaften
University of Erlangen-Nürnberg
West Germany

Translated by
Ursula E. Wolff

 1980

ACADEMIC PRESS

A Subsidiary of Harcourt Brace Jovanovich, Publishers

New York   London   Toronto   Sydney   San Francisco

ACADEMIC PRESS, INC.
111 Fifth Avenue, New York, New York 10003

United Kingdom Edition published by
ACADEMIC PRESS, INC. (LONDON) LTD.
24/28 Oval Road, London NW1 7DX

Library of Congress Cataloging in Publication Data

Heimendahl, Manfred von.
    Electron microscopy of materials, an introduction.

    (Material science and technology series)
    Translation of Einfuhrung in die Elektronenmikroskopie

    Bibliography: p.
    Includes index.
    1. Electron microscopy. I. Title. II. Series.
QH212.E4H4513   530.1'1   79–6810
ISBN 0–12–725150–2

PRINTED IN THE UNITED STATES OF AMERICA

80 81 82 83   9 8 7 6 5 4 3 2 1

Electron Microscopy of Materials translated from
the original German edition entitled Einführung
in die Elektronenmikroskopie, copyright © 1970
by Friedr. Vieweg & Sohn GmbH, Verlag,
Braunschweig.

Contents

Preface to the German Edition   viii
Preface to the English Edition   x
List of Materials and Abbreviations   xi

1. Fundamentals and Techniques

1.1   Historical Introduction   1
1.2   Electron Beams as Waves   2
1.3   Lenses for Electron Beams   4
1.4   Structure of Transmission Electron Microscopes   6
1.5   Lens Defects and Resolution   10
1.6   Mechanism of Image Formation; Contrast   15
      1.6.1   Amorphous Specimens   15
      1.6.2   Crystalline Specimens   19
1.7   High Voltage Electron Microscopy   29
1.8   Surface Electron Microscopes   32
1.9   Scanning Transmission Electron Microscopy (STEM) and Energy-
      Dispersive X-Ray Analysis (Elemental Microanalysis)   37
      References   44

2. Preparation

2.1   Production of Thin Foils by Electrolytic Polishing   47
      2.1.1   Prethinning to a Thickness of 0.15 mm   47
      2.1.2   Window Method   49
      2.1.3   Bollmann Technique   56
      2.1.4   Jet-Polishing Methods   58
      2.1.5   Preparation of Wires   61
2.2   Production of Transparent Samples by Other Procedures   62
      2.2.1   Mechanical Procedures (Hammering, Cleaving, Cutting, Ion
              Thinning)   63
      2.2.2   Chemical Methods   71
2.3   Replica Technique for Surfaces   74
      2.3.1   Single-Stage Techniques (Plastic, Carbon, Oxide Replicas) 74

2.3.2   Two-Stage Techniques (Technovit, Triafol)   80
2.3.3   Extraction Replicas   86
2.4   Preparation of Powders   89
Appendix   91
References   92

3. Electron Diffraction

3.1   Fundamentals, Comparison with X-Ray Diffraction   95
3.2   Debye–Scherrer Patterns, Standardization   99
3.3   Reciprocal Lattice   102
3.4   Construction of Simple Diffraction Patterns   105
3.5   Method of R_n Ratios   108
3.6   General Case of Indexing Single-Crystal Diffraction Patterns 113
3.7   Correlation of Image and Diffraction Pattern: Magnetic Rotation 116
3.8   Determination of Directions and Planes; Trace Analysis   118
3.9   Foil Thickness Determination by the Trace Method   120
3.10 Uniqueness (Unambiguity) of Orientation Determination   127
3.10.1   The 180° Ambiguity   127
3.10.2   Coincidence Ambiguity   129
3.11 Kikuchi Lines   130
References   139

4. Contrast Theory and Applications

4.1   Kinematical and Dynamical Theory; Concepts   141
4.2   Derivation of the Basic Equation of the Kinematical Theory   144
4.3   Ewald Sphere   145
4.4   Amplitude–Phase Diagram   148
4.5   Application of the Basic Equation to Ideal Crystals   151
4.5.1   Deviation Parameter s and its Experimental Determination   151
4.5.2   Crystal in the Shape of a Brick-Shaped Parallelepiped   154
4.5.3   Extinction Distance   161
4.5.4   Application Examples: Wedge Fringes, Bend Contours   163
4.5.5   Thickness Determination Based on Section 4.5.4   169
4.6   Application of the Basic Equation to Real Crystals   171
4.6.1   Dislocations   173
4.6.1.1   Derivation of the contrast of a screw dislocation   173
4.6.1.2   The $g \cdot b$ criterion   177
4.6.1.3   Edge dislocations   178
4.6.1.4   Double contrast and more complicated cases of
          dislocation contrast   182
4.6.1.5   Determination of Burgers vectors b   183
4.6.1.6   Determination of dislocation densities   185
4.6.1.7   Examples of dislocation configurations   188
4.6.2   Stacking Faults   190

4.7   Analysis of Twin Structures   194
4.8   Contrast of Precipitates   196
        4.8.1   Guinier–Preston Zones (GP Zones)   198
        4.8.2   Strain Contrast   199
        4.8.3   Criteria According to Ashby and Brown   200
        4.8.4   Metastable and Stable Phases   202
        4.8.5   Contrast Reversal at Thickness Fringes   203
4.9   Moiré Patterns   205
4.10   Ordered Structures   209
4.11   Magnetic Materials   211
4.12   Methods for Improving the Accuracy of Orientation
        Determination and for Obtaining Unambiguity   212
4.13   Concluding Remarks   216
        References   217

Author Index   219
Subject Index   223

Preface to the German Edition

This book is aimed at students and at all those practicing physicists, engineers, and technicians who either have to familiarize themselves with electron microscopic examination techniques for materials, or who want to use the results of electron microscopic investigations in their work. For this growing group the book may serve as an *elementary primer*. It should enable them to learn the more important preparation techniques, to evaluate diffraction patterns with confidence, and to apply simple (kinematical) contrast theories to the interpretation of transmission electron micrographs. Examples of techniques are discussed in detail; they are at the same time examples of electron microscopic imaging of various materials.

This introduction is conceived of as a short textbook which should accompany and assist readers in their practical work. The subject matter is presented in the form of examples which should be grasped easily. The only *prerequisites* are those which generally are taught during undergraduate study of physics or engineering (e.g., fundamentals of light optics and of materials science or metallurgy). In addition, it is useful to have some knowledge of x-ray diffraction as well as of the pertinent crystallographic fundamentals (Miller indices, stereographic projection).

With a few exceptions, all procedures described in this book have been personally tried and employed. They are easily reproducible and are also suitable for *student experiments*.

If more complicated problems arise, an extensive bibliography is available at the end of each chapter. The works cited there also contain many original contributions. Generally, these have not been quoted in this book in conformance with its character as a textbook. Recent original publications have been cited in only a few cases, also as sources of some of the illustrations used.

The book originated from lectures and laboratory courses given by the author from 1966 to 1969 at the Technical University, Clausthal, and at the Institute for Materials Science of the University of Erlangen-Nürnberg (both in West Germany).

The help of co-workers and the encouragement of the directors of both

Institutes, Professors G. Wassermann and B. Ilschner, are gratefully acknowledged. Special thanks are due to Chr. Pommerehne, Clausthal, and D. Puppel, Erlangen, for their outstanding technical preparations. Most of the electron micrographs in this book for which no other sources are cited were produced in collaboration with them. The author also wishes to thank those colleagues and companies who graciously supplied additional illustrations.

Professor E. Brüche, Dr. W. Pitsch, Düsseldorf, and a few colleagues at the Institute for Materials Science of the University of Erlangen-Nürnberg have reviewed individual chapters. Their efforts and valuable suggestions are greatly appreciated.

M. VON HEIMENDAHL

Erlangen, September 1969

Preface to the English Edition

For the English translation, this book has been slightly revised and misprints have been corrected. Sixteen illustrations have been added and three have been improved. Sections 2.1.4 and 3.10 have been rewritten and a new Section 1.9 has been added to include the advances made in these fields.

The author is particularly grateful to Ms. Ursula E. Wolff for her careful and conscientious translation of the book, including some valuable suggestions, and for proofreading. Thanks are extended to Mr. H. Giese and Dr. E. Tenckhoff for valuable discussions and a critical reading of Section 1.9.

M. VON HEIMENDAHL

Erlangen, October 1979

List of Materials and Abbreviations

List of Materials Used in Figures as Examples for Preparation or Analysis Procedures

Metals					Nonmetals	
	Figure		Figure			Figure
Al	1.4, 2.21, 3.5b, 3.17,	CoNiCrFe	4.31		Al_2O_3	2.23B
	4.13, 4.15, 4.24	Fe	4.44		barium titanate	2.25
Al–0.2%Au	4.36, 4.40	FeCo	4.42		carbon film	1.5a
Al–4%Cu	1.10, 3.12,	Fe–4%Mo	4.28		glass (Cu–ruby)	2.14
	3.13, 3.14, 4.33, 4.35	Steel			LiF	2.19
Al–1%Mg	2.13	pearlite + martensite	1.15b		$MgO–Cr_2O_3$	2.16
AlCuMgPb	1.13	pearlite	2.18		mica	2.10
AlZnMgl	1.18, 4.26a,	stainless	4.32		Mn nodule powder	2.29
	4.37	10CrMo9 10	2.27B		nylon fiber	1.17b
amorphous metal		Mo	1.15a		paper	2.20
FeNiCrPB	1.21	Ni	4.27		plastic (PVC)	2.15
Au	1.5b	TD–Ni (2%ThO$_2$)	4.14,		quartz porcelain	2.26
Au (gold leaf)	2.9, 3.3d		4.29		Steatite	2.24
Cu	2.8, 4.25	Nimonic	1.17a		TlCl	3.3
Cu–3%Co	2.12	Permalloy	4.26b			

Abbreviations Used in This Book

EM	electron microscopy, electron microscope
TEM	transmission electron microscopy
SEM	scanning electron microscopy
STEM	scanning transmission electron microscopy
SAD	selected area diffraction
APD	amplitude-phase-diagram
rel	reciprocal lattice (point, plane, etc.)
bcc	body-centered cubic
fcc	face-centered cubic

Seeing is Believing

1. Fundamentals and Techniques

1.1 Historical Introduction

In 1925–1927, Busch discovered that a rotationally symmetric, inhomogeneous magnetic field could be conceived of as a *lens* for an electron beam, in analogy to a glass lens for a light beam. Similar lenslike characteristics for electron beams were also found for electrically charged slit or hole apertures. The idea to use electron beams for producing *enlarged images* was first carried out in 1932 by two independent research groups: Knoll and Ruska (Technical University of Berlin) who produced a magnetic-type electron microscope, and Brüche and Johannson (AEG-Research Institute of Berlin) who produced an electrostatic type. The resolution limit of the light microscope was exceeded a few years later by Krause.

Further development of the electron microscope, still on a laboratory scale, was made possible through the work of numerous scientists in industrial and university institutes. Those in Germany who took a decisive part, in addition to those already mentioned, included v. Ardenne, Boersch, v. Borries, and Mahl. Perfected electron microscopes (EMs) could be mass produced from the beginning of the 1950s. Since then, their development has steadily progressed, mainly in the direction of increasingly better resolving power and of increasingly higher accelerating voltages.

In addition to medical–biological applications, the electron microscope was used from the beginning for the investigation of materials. In those early times such investigations depended on the laborious and indirect replica technique. Then a crucial breakthrough of special significance occurred in 1957–1958. During that period, Hirsch and co-workers (Cambridge, England) first made rapidly increasing use of the possibility of electrolytically thinning metal foils and observing these with transmitted electrons. Thus direct pictures of the interior of matter were obtained. In 1949, Heidenreich for the first time had observed thinned aluminum foils with a transmission electron microscope. However, it was several years before commercial electron microscopes with double condensers and 100 kV accelerating voltages had become sufficiently numerous. Only these possess the light intensity for the observation of thin (50–300-nm-thick) crystalline samples.

It can indeed be said, without exaggeration, that since then an entire new world has been opened for the investigation of metals as well as for many nonmetallic materials. Various lattice defects, which until then had only been theoretically described or indirectly demonstrated, could finally be made directly visible. Examples are dislocations, stacking faults, very small precipitated second phases, grain boundary details, and the determination of the crystallographic orientation of small lattice regions or of particles of an order of magnitude of 1 μm or less. These advances have been reflected since about 1960 in an avalanche of publications, too much for one person to digest fully. During the years 1957–1969, the number of publications in the field of transmission electron microscopy (TEM) of metals and alloys alone was close to 3000. Nevertheless, the total volume of matter investigated until now with the electron microscope, in the entire world, is less than 1 mm^3! This astonishing fact is entirely due to the high magnifications of the electron microscope.

There are several thousand electron microscopes in the world. Assuming 10,000 microscopes as an upper limit, and taking the average number of pictures produced as 20,000 per microscope, 2×10^8 micrographs would have been taken. At an average magnification of $20,000\times$, a picture size of 6×9 cm and a sample thickness of about 0.1 μm, the volume examined per picture would be 1.4 μm^3 or the total volume 2.8×10^8 μm^3, i.e., <0.3 mm^3. (This estimate is due to Swann.)

For more comprehensive books on electron microscopy, see [1] to [15b].

1.2. Electron Beams as Waves

The goal of all microscopy is to obtain enlarged pictures of objects with the best possible *resolution*. Resolution is defined as the distance g of two object details just separable (resolvable) from one another. In all of optics, the *Abbé theory* is applicable to the mechanisms of image formation and resolution. If the object is assumed to be a lattice of lattice constant g (representing arbitrary object details), this theory states the following: To form an image of this object, as many as possible of the *diffraction maxima* (Fraunhofer diffraction), but at least the first diffraction maximum, must pass through the objective lens or, more specifically, through the aperture, in order to produce, by combination with the main beam, an image behind the lens. The equation for the resolution g according to Abbé's theory follows from this condition (see Fig. 1.1). If α is the angle between incident and deflected beam, then the first maximum occurs at that angle α for which the rays from adjacent lattice openings have a phase difference of one wavelength λ. The smaller g is, the larger α. For

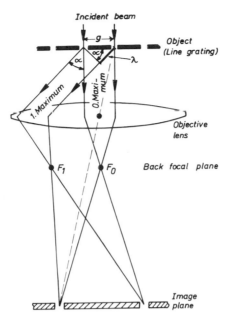

Fig. 1.1. Ray path diagram defining resolving power. Fraunhofer diffraction at a line grating (lattice constant g). In the back focal plane the zero-order beam is focused in focal point F_0, the 1st-order beam in F_1, etc. All beams of different orders are recombined in the image plane to form the image.

this condition, it follows from Fig. 1.1 that: $g = \lambda/\sin \alpha$. A better theoretical treatment for the resolution yields a numerical factor of about 0.6. Thus

$$g = 0.6\lambda/\sin \alpha. \tag{1.1}$$

(This numerical factor of <1 is physically based on the fact that only *part* of the first diffraction maximum is necessary for image formation.) Since the rays deflected by the angle α must be able to enter the front lens of the objective, α is also called the *opening angle* or *aperture angle* of the objective lens.

The limit of resolution according to Eq. (1.1) depends on the wave nature of the imaging light (diffraction phenomena). Equation (1.1) assumes an ideal lens. If lens defects are present, the resolution is correspondingly lower (Section 1.5).

In order to obtain the best possible resolution (i.e., smallest g), the largest possible aperture α has to be used. In light optics, by elimination or improvement of lens defects, one can work with apertures of the order of magnitude 1. Therefore, the resolution is limited by the *wavelength* of

Table 1.1 **Wavelengths and Velocities of Electrons of Various Energies**

U (kV)	λ (nm)	v (10^{10} cm/s)
40	0.00601	1.1216
60	0.00487	1.338
80	0.00418	1.506
100	0.00370	1.644
200	0.00251	2.079
500	0.00142	2.587
1000	0.00087	2.822

the light used (400–800 nm). To obtain fundamentally better resolutions (and higher useful magnifications), waves suitable for image forming purposes with *substantially shorter wavelengths* had to be found.

In 1924, de Broglie had postulated that—considering the dual nature of waves and particles—a wavelength λ can be associated with particle beams, including electron beams. For electrons of mass m and velocity v, the wavelength is, according to de Broglie, $\lambda = h/mv$, h being Planck's constant.

When electrons are accelerated through a potential U, they have an energy of $Ue = \frac{1}{2}mv^2$ (e is the electron charge). With the above equation for λ, it follows that the relation between wavelength and accelerating voltage is

$$\lambda = h/(2mUe)^{1/2} \quad \text{or} \quad \lambda = [1.5/U(\text{volt})]^{1/2} \quad [\text{nm}] \quad (1.2)$$

if the numerical values for h, m, and e are substituted. When v becomes comparable with c, the speed of light, the right-hand side of Eq. (1.2) has to be multiplied by the relativistic correction factor $(1 + Ue/2m_0c^2)^{-1/2}$ and m has to be replaced by m_0, the rest mass of the electron. Table 1.1 gives the wavelengths thus derived and also the velocities for monoenergetic electrons of the more prevalent accelerating voltages for the EM (after Hirsch et al. [1] and Thomas et al. [2]). This table shows that the wavelength of electron waves with energies corresponding to accelerating voltages in the range 40–100 kV is about *five orders of magnitude smaller than the wavelength of visible light*.

1.3. Lenses for Electron Beams

In order to obtain an image, efficient enlarging *lenses* are required (for an exception see Section 1.8). It turns out that electron beams can be deflected and focused by electrostatic or electromagnetic lenses in complete analogy to the focusing of light waves by glass lenses. Today, elec-

trostatic lenses play only a comparatively minor role and therefore will
not be discussed in more detail.

Figure 1.2 shows schematically, side by side, a glass lens and an electromagnetic lens. The latter consists of a current-carrying coil surrounded
by iron. The magnetic field lines emerge from only a small (air or brass)
gap in the iron case. In this region, therefore, the field is very concentrated, intense, and, moreover, strongly inhomogeneous (Fig. 1.2c).

If electrons, with charge e and velocity v, approach the lens nearly
parallel to the optical axis (or at a small angle to it), the force **F**, exerted
by the magnetic field **B** on an electron at any given point, is the Lorentz
force

$$\mathbf{F} = e[\mathbf{v} \times \mathbf{B}].$$

The force thus acts normal to the direction of the magnetic field and
normal to the velocity of the electrons, i.e., normal to the plane of the
drawing. The force **F**, therefore, forces the electrons out of the axial
direction and into a helical path, the axis of the helix being the optical
axis. Moreover, it is essential that the magnetic field exhibit the strong
inhomogeneity already mentioned. Thus, the electrons, passing through
the field, are *deflected toward the optical axis,* as can be calculated in
detail [3]. Accordingly, the effect of the coil on the electron beam is comparable to the refraction of a glass lens acting on light rays. In both cases,

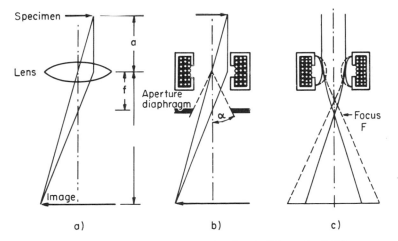

Fig. 1.2. Lenses in light and electron optics. a) Glass lens in light optics, b) Electromagnetic
lens in electron optics. α aperture angle. c) as b). Magnification due to weak coil current,
i.e., low magnetic field ———, or strong coil current, i.e., high magnetic field strength
– – –.

the same laws of geometric optics apply. Thus, two well-known formulas exist between object distance a, image distance b, focal distance f, and magnification M:

$$1/a + 1/b = 1/f \quad \text{and} \quad M = b/a.$$

A beam approaching from a great distance and traveling parallel to the axis is focused at the focal point F (see Fig. 1.2c). This figure also illustrates that the focal distance, and thereby the magnification, will vary with varying coil currents (refractive power). In electron optics only converging lenses exist, no diverging lenses.

Electron lenses are afflicted with the same types of *lens defects* as glass lenses (namely: astigmatism, chromatic aberration, spherical aberration, etc.), but their degree of importance is different. Also, the lens defects of electron lenses are *considerably greater* than those of glass lenses. In principle, all lens defects decrease with decreasing aperture angle α. This angle is determined by a diaphragm, the *aperture diaphragm*, placed in the focal plane behind the lens, see Fig. 1.2b. In approximation tan α = diaphragm radius/focal distance. While in light optics, as mentioned, aperture angles of the order of magnitude 1 can be used, in electron optics, because of the much larger lens defects, α must be limited to a very small value (about 1/100 in radians, corresponding to 1/2°). Thus, although the wavelength of electrons is shorter than that of light by 5 orders of magnitude, the large lens defects prevent corresponding improvements in resolution [Eq. (1.1)]. From Eq. (1.1) one would expect that with sin α $\approx \alpha = 1/100$, the resolution would be about 100 times the wavelength, i.e., for 100 kV electrons (see Table 1.1) about 0.3 nm. This value is, indeed, the lower limit of resolution obtainable with present day electron microscopes.

In Section 1.5, lens defects and resolution will be discussed in greater detail. First, however, it is useful to treat the various configurations of electron microscopes possible since the development of electron lenses. This is done first in summary form, to obtain an overview and to characterize the most important individual parts of the microscope. More detailed descriptions will be given later.

1.4. Structure of Transmission Electron Microscopes

A fundamental difference exists between transmission and surface electron microscopes. For the former, the sample must be so thin that it is transparent to electrons while for the latter, solid bulk samples can be used (Section 1.8).

Most electron microscopes are constructed similar, in principle, to light microscopes, i.e., they are composed of a series of enlarging lenses.

The basic design of a transmission electron microscope (TEM) is discussed with the aid of Fig. 1.3. A hairpin cathode (tungsten),[1] heated to incandescence, emits electrons which are accelerated by the anode and collimated ("cross-over") by the Wehnelt cylinder at the anode. The divergent electron beam is focused exactly on the object or specimen by the condenser lens. (In practice, usually two condenser lenses are present, a double condenser, used in series to intensify the effect). Thus, with a double condenser, the electrons leaving the tip of the hairpin cathode are focused at the specimen position to a very small spot of only 3–5 μm in diameter. This is an essential condition for high light intensity on the image screen of a transmission electron microscope.

If the object is crystalline, then, in addition to the primary beam, *interference* maxima are formed behind the sample as a result of *Bragg diffraction* in the crystal lattice of the specimen, analogous to x-ray diffraction (for more details see Section 1.6.2 and Chapter 3). For metal samples, the Bragg diffraction angles are so large that the diffracted beams are intercepted by the objective aperture. The latter is located in the back focal plane of the objective lens. There, each beam deflected by Bragg diffraction is focused to a spot. The entirety of these spots (only two are shown in Fig. 1.3) is the Bragg *diffraction pattern* (Chapter 3).

With the beams passing through the objective aperture, the objective lens forms the single-stage magnified image of the sample. In the plane of this "first intermediate image" the so-called selector (or intermediate) aperture is located. The intermediate lens and projector lens[2] successively enlarge the first intermediate image two more times. The total magnification, again as in light optics, is obtained by multiplication of the individual enlargements of the series of lenses. The three-stage magnified image is visible on the fluorescent, final image screen or, when the screen is lifted, is recorded on the photographic plate.

The following numerical data[3] may give a better idea.

- Acceleration voltages selectable at 40, 60, 80, or 100 kV (see, however, Section 1.7); the anode is grounded.

---

[1] Other electron sources are LaB$_6$ and field emission cathodes. Although these are much more expensive than the regular tungsten cathodes, they have gained popularity because they are considerably brighter (a great advantage at high magnifications).

[2] These are occasionally also called 1st and 2nd projector lenses.

[3] Valid for the "Elmiskop I" of Siemens AG; in the following, this widely available type of microscope will be used as a basis for numerical data.

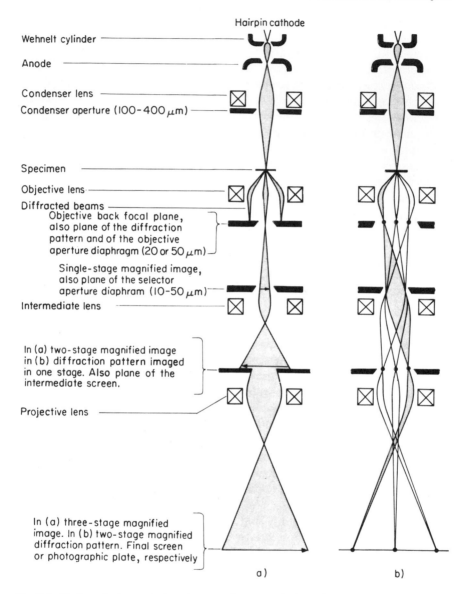

Fig. 1.3. Design of a transmission electron microscope (schematic). a) Ray path for bright field imaging (three-stage magnification). b) Ray path after Boersch for selected area diffraction (SAD). The excitation of the intermediate lens is weaker in b) than in a); thus, the primary diffraction pattern is imaged by the intermediate lens instead of the first-stage sample image.

- Cathode heating, 2–3 V: electron current (beam current) 0–60 μA adjustable by an auxiliary bias voltage at the Wehnelt cylinder.
- Magnification by the objective lens alone, 200×; (however, see Section 1.6.2).
- Magnification through the objective and intermediate lens, 2600×.
- Useful electron-optic magnification through all three lenses 8000–200,000×, adjustable by varying the intermediate lens current and by using different projector pole pieces.
- Cathode–anode separation, approximately 13 mm.
- Objective focal distance, 2.8 mm.
- Total microscope length, 1.25 m.
- Vacuum, 10^{-4}–10^{-5} Torr; the specimen chamber and the chamber for the photographic plates can be separated from the vacuum by individual airlocks for quick exchange of samples or photographic plates.

The variable magnification can be adjusted over a large range by means of the variable intermediate lens excitation. The principle is as follows.

Since the projector lens has a constant magnification, the position of the second stage enlarged image (see Fig. 1.3a) in the EM is fixed. Consequently, the image distance b_I of the intermediate lens is likewise constant. According to

$$\frac{1}{f_I} = \frac{1}{a_I} + \frac{1}{b_I} \quad \text{and} \quad M_I = \frac{b_I}{a_I},$$

where M is the magnification of the intermediate lens, see equations from Section 1.3, the magnification of the intermediate lens can be varied by changing its excitation (i.e., by varying its focal distance f_I). The respective object distance a_I satisfying the lens equation thus becomes the dependent variable. For each magnification, the distance a_I defines the position of the intermediate image (i.e., the image magnified by the objective lens alone). Consequently, b_O (image distance of the objective) is also determined. Since, on the other hand, the object distance of the objective, i.e., the specimen position, is structurally fixed, the focal distance f_O, according to

$$\frac{1}{f_O} = \frac{1}{a_O} + \frac{1}{b_O},$$

must be adjusted as a dependent variable in order to satisfy the lens equation for the objective. This is the procedure of focusing. The objective is a very strongly magnifying lens. Therefore, a_O is only a little larger than f_O, and the variations of f_O for focusing are very slight (order of magnitude 0.1–10 μm). Thus, one still can talk of "the" focal length of the objective. On the other hand, from these considerations it can be seen that even small deviations from the normal sample position can strongly affect the magnification, because focusing of the objective lens also changes its magnification. This must be taken into account for quantitative magnification determinations, in case the position of the specimen cannot be precisely controlled.

Figure 1.4 illustrates with a single sample area the effect of some basic magnification steps (transmission micrograph of an electrolytically thinned Al foil, Section 2.1). As a result of the helical path of the electrons in the EM (see preceding section), the image is rotated when the magnification is changed (magnetic rotation, Section 3.7).

Further details will be discussed in the following chapters. Always refer to the detailed operating instructions or to the prospectus of the respective microscope manufacturers which contain necessary supplemental information. This book is not intended as a substitute. For this reason this book also contains no illustrations of commercial instruments. This applies likewise to the numerous, often very worthwhile, attachments, in particular, the special object cartridges (sample holders) which are used to tilt the sample about defined axes perpendicular to the beam, to heat, cool, or stretch them elastically–plastically, or to perform other similar manipulations during observation.

1.5. Lens Defects and Resolution

As mentioned in Section 1.3, the lens defects of electromagnetic lenses present a considerable technical problem. The following are of particular importance:

a) astigmatism (a point will not be imaged as a point, but rather as a small line),
b) chromatic aberration (different refractive indices for different wavelengths λ),
c) spherical aberration (the focal length f of the outer lens zone is smaller than that of the inner zone).

Astigmatism occurs because the magnetic field of the lens, for technical reasons, can never be produced ideally rotationally symmetric (noncircular drilling, contamination on the edge of the pole piece, etc.). However, this defect can be *compensated* for. A "stigmator", which consists of movable soft iron pieces, is located in the objective lens of older microscopes. Modern electron microscopes employ, instead, a small system

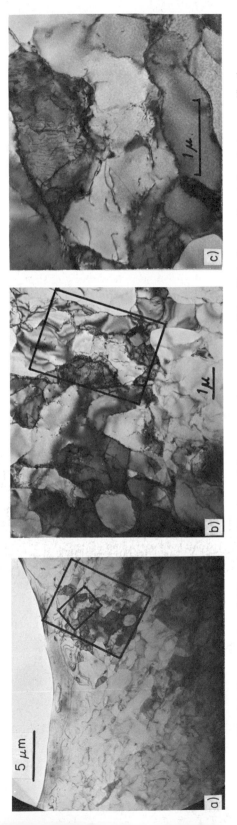

Fig. 1.4. Demonstration of different magnifications of the same specimen area and of the effect of magnetic rotation during magnification changes. The rectangular areas used for the higher magnification steps are marked. *Material:* Al 99.99, 50% cold-rolled (single dislocations, dislocations forming cell structure, and subgrain structure.) Electrolytically thinned foil (for preparation see Section 2.1). a) 2600×, only objective and intermediate lens switched on; b) 8000×, objective, intermediate, and projective lens (with pole piece 1) used; c) 20,000×, as b) but pole piece 3 used in the projective lens (Siemens Elmiskop IA. 100 kV).

of auxiliary coils in pairs. These coils are suitably excited for astigmatism correction. Thus, in practice, astigmatism does not limit resolution of EMs.

The same applies to *chromatic aberration*. By suitable *stabilization of the high voltage* (residual beam voltage variations attainable are $\leq 1 \times 10^{-5}$ at 100 kV in 1 min), sufficiently monoenergetic electrons can be obtained. *Lens currents* also must be well stabilized, so that the focal lengths do not change.

There remains the *spherical aberration* which cannot be compensated for by dispersion lenses as in light optics since, of course, no such lenses exist in electron optics. *In practice, therefore, spherical aberration determines the attainable resolution.* If g_s is the resolution limit due only to spherical aberration, then the diameter g_s of "the disk of confusion" is $g_s = C_s\alpha^3$ where C_s is the coefficient of spherical aberration and α is the aperture angle. As mentioned in Section 1.3, α must be kept *as small as possible,* in order to sufficiently reduce the lens defect g_s. On the other hand, according to Eq. (1.1) good resolution requires, in principle, an aperture α *as large as possible.* Since the resolution limit given by Eq. (1.1) is determined by the wave nature of the imaging rays (diffraction phenomena instead of geometric optics), the quantity $g_b = 0.6\lambda/\alpha$, defined by Eq. (1.1), is also designated as *diffraction error*.

The resolution terms are approximately additive and the *effective resolution* g resulting from the spherical aberration g_s and the diffraction error g_b [3] is

$$g = g_s + g_b = C_s\alpha^3 + (0.6\lambda/\alpha).$$

The minimum possible g resulting from the variation of α can be calculated through differentiation of this equation. This minimum is the *theoretical resolution limit of the* EM. The corresponding optimum value α_{opt} is obtained by solving the minimum problem:

$$\alpha_{opt} = (0.6\lambda/3\,C_s)^{1/4}.$$

Substituting this value into the last equation, one obtains for the *theoretical resolution limit*

$$g_{min} = A\,(C_s\lambda^3)^{1/4}, A = 1.2. \tag{1.3}$$

The coefficient of spherical aberration C_s for EM-objective lenses is of the order of 1 mm, according to the current state of technology. It follows that $\alpha_{opt} = 5 \times 10^{-3}$ for $\lambda = 0.0037$ nm (100 kV electrons). For a focal length of $f = 2800$ μm, the optimal diameter of the objective aperture is $2\alpha f = 28$ μm. The objective apertures actually used in the EM (see Fig. 1.3) have, therefore, diameters of 20, 30, or 50 μm. A 50 μm aperture corresponds to $\alpha \approx 1/100$.

A more exact, wave-mechanical calculation of the theoretical resolution limit yields a value of 0.43 for the factor A in Eq. (1.3). With this A, for 100 kV electrons and $C_s = 1$ mm, *the theoretical resolution limit* is $g_{min} = 0.2$ nm.

The resolution obtainable in practice is a function of the expense or the care with which one reduces the defect sources mentioned. Modern mass produced electron microscopes guarantee an obtainable point-to-point resolution of 0.2–0.5 nm. With the best available instruments and by taking special care, it has been possible to attain a resolution of 0.13–0.2 nm. The above calculated theoretical resolution limit thus has been attained.

When evaluating an instrument one should consider, however, the following facts:

a) In practice, the highest resolving power often cannot be fully utilized because only a few objects exhibit sufficient *contrast* of object details (see Section 1.6) on such a small scale. (The contrast diminishes with increasing magnification!) In routine operation, samples with medium or moderate contrast often allow a practical utilization of the resolution only down to 3–2 nm. Although optimal resolution is a distinctive feature of an EM, one should not place too much emphasis on it. In practice, many other features such as reliability, easy serviceability (safety circuits against faulty operation), light intensity, etc., are often more important criteria for the evaluation of an EM.

b) When specifying or evaluating the resolving power of a microscope, one must be sure to distinguish between *point resolution* and *line resolution*. It is easier to resolve two adjacent parallel lines than two points with the same separation. This is illustrated in Fig. 1.5. Figure 1.5a shows an often utilized test sample for resolution testing, namely a carbon supporting film in which two points with a separation of 0.5 nm have a separation in the micrograph of 0.5 mm at 1,000,000 × magnification (electron optically 210,000 ×, then further photographically enlarged). Figure 1.5b is an example of line resolution: it has become possible to image in the TEM lattice planes from metal crystals, here the (111) planes of a gold crystal lattice, with a lattice spacing d of 0.235 nm. Special imaging techniques must be used for such samples in order to actually reach the extreme limit of presently possible resolution (see Section 1.6.2). More recently, several investigators [21–23] have even succeeded in direct imaging of lattice defects in the lattice planes.

c) Finally, a comment should be made regarding the often raised ques-

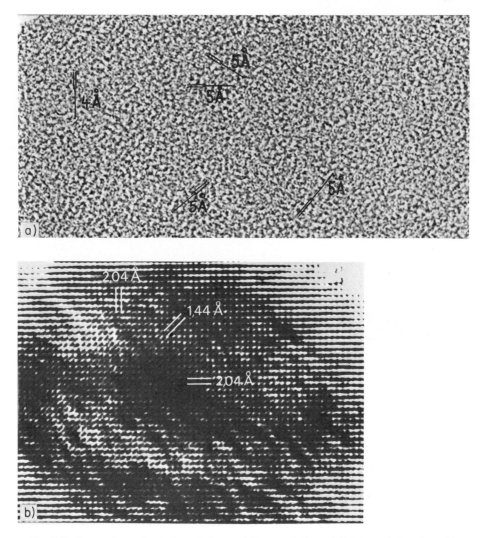

Fig. 1.5. Comparison of point resolution and line resolution. a) Point resolution in a thin carbon film, 1,000,000×. Illuminating aperture 4×10^{-4} and objective aperture 1.3×10^{-2}, 100 kV. b) Line resolution in a thin gold foil (epitaxially grown), (200), (020), and (220) lattice planes. Bright field imaging, 125 kV, oblique illumination (see Section 1.6.2). (Both micrographs: Courtesy Siemens AG., Germany.)

tion, whether one could hope to make individual atoms visible in the TEM. With the now attainable resolution of 0.2–0.3 nm, this appears, theoretically, to be within reach. Practically, however, it is extremely difficult. Besides difficulties due to image theory, in practice, such resolutions cannot be obtained for materials specimens, since the thinnest producible specimen foils are still some 10 nm thick. Consequently, the electron beam traversing the specimen "sees" at least about 100 atom layers, *one above another,* and therefore cannot resolve single atoms. (For the imaging of lattice planes, this argument does not hold, provided the lattice planes are oriented exactly parallel to the beam.) The question of the imaging of atoms will probably remain, therefore, a domain of the field ion microscope, which in many cases has already solved this interesting problem. Recently, however, it has become possible to resolve single heavy atoms or atom groups (e.g., Hg, Pt) in a completely different situation. If such heavy atoms lie isolated on a supporting thin organic film, they can be resolved by high resolution TEM.

1.6. Mechanism of Image Formation; Contrast

In the previous discussion, certain analogies to the conditions in light optics were emphasized (applicability of the laws of geometric optics and the Abbé theory). If the sample in the EM would, for example, consist of a line grating made of lead, with bars so thick that they completely absorb the electrons, this analogy would remain valid (shadow image production).

Actual electron microscope specimens, however, are different. As a result of the preparation, they are so thin that they are more or less transparent to electrons. The processes during passage of the electrons through the sample, therefore, must be considered in more detail. It will be shown that in spite of common physical principles, several important differences exist between the mechanism of image formation with electrons and with light.

In addition to the fundamentals mentioned, the actual *brightness of a particular image point* in light optics depends mainly on the *absorption* of the light in different sample areas. With the EM, because of the transparence of samples for electrons, absorption plays only a minor role and *deflection mechanisms* (scattering and diffraction) are mainly responsible for image contrast.

It is practical to differentiate between amorphous and crystalline specimens.

1.6.1. Amorphous Specimens

During passage of an electron wave through a specimen, *scattering* of the electrons occurs at the atoms in this specimen: This means that—

purely phenomenologically—a certain portion of the incident electrons is *deflected* (scattered) in directions different from the primary beam direction, Fig. 1.6a. This scattering of electrons occurs over all angles, but is strongest in the "forward" direction, i.e., for scattering angles near zero degree. This fact of decreasing scattering intensity with increasing angle of scatter is indicated by the different arrow lengths in Fig. 1.6a and is schematically drawn in Fig. 1.6b.

The *physical cause of scattering* is the electric charge of the electrons and their interaction with the atomic nucleus. The deflection of the electrons in the beam results from the Coulomb field of the atomic nucleus. Since the mass of the nucleus is much larger than that of an electron, the nucleus practically does not change its position. *No energy transfer worth mentioning* occurs by the scattering process. Therefore, this type of scattering is also referred to as *elastic scattering* of the electron.

Quite different is the scattering process in which a beam electron encounters an electron in the atomic shell. Due to the identical masses, a more or less extensive energy transfer occurs (depending on the scattering angle). Therefore, this process is called *inelastic scattering*. In *electron microscopy*, basically both types of scattering occur; in practice, however, elastic scattering at the nucleus predominates. The scattering of *x-rays*, on the other hand, almost exclusively is due to scattering at the atomic shell electrons.

Before the effect of elastic scattering upon image formation is discussed, it should be noted that by inelastic scattering with energy transfer, the electron can be decelerated down to thermal energy and can no longer leave the sample. It remains as a lattice electron. In this case, *absorption* of electrons takes place in the sample and a certain amount of *heat* is created. In addition *x rays* are created in the EM by collisions between the beam electrons and the atoms. Since in the EM relatively high electron energies (100 keV) and small sample thicknesses (approximately 0.1 μm) are used, both effects (absorption and x-ray production) play only a minor

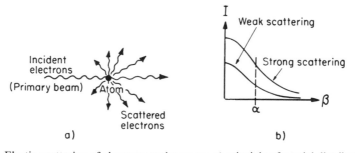

Fig. 1.6. Elastic scattering of electrons at the atoms, a) principle of spatial distribution and b) scattering intensity vs. scattering angle β (schematically).

role. Under the stated conditions, most electrons leave the sample again, without being absorbed.

It should, however, be mentioned that, due to the moderate fraction of electrons absorbed, measurable heating of the specimen can result. This depends strongly on the magnitude of the beam current, as well as on the thickness and heat conductivity of the specimen. As a result, quite different values are reached (from a few degrees up to the order of magnitude of 100°C; for more details see Hirsch *et al.* [1]). Although the x rays originating in the sample are hard (100 kV), they have such weak intensity that absorption in the wall of the EM gives sufficient radiation protection to the observer. In contrast, x rays produced on the fluorescent screen are much more penetrating. Therefore the observation windows must be made of lead glass for shielding.

As an aid to understanding the effect of scattering upon image formation, Fig. 1.7 shows the essential region around the objective lens (enlarged from Fig. 1.3).[4] The primary beam impinging on the sample from above can be viewed as being a nearly parallel beam in this "enlarged portion" of the beam path. The sample is located very near the upper focal point (slightly above), since this will result in a greatly enlarged image. The objective aperture is situated in the lower focal plane. With this and all similar diagrams it must be kept in mind that the angles are drawn greatly exaggerated. (In reality, the aperture angle is only 0.5°, see the numerical data in the figure caption.)

Consider, for example, a specimen consisting of very light atoms A with an inclusion consisting of atoms B with a higher atomic number Z. The area of the inclusion is drawn as a dark cross section in Fig. 1.7. While the electrons can penetrate the areas A of the specimen almost without hindrance (scattering negligible), the electrons directed at B are deflected comparatively much more strongly by scattering because B consists of heavier atoms of greater charge. These scattered rays are indicated in Fig. 1.7. One remembers that the objective aperture is very narrow ($\alpha = 1/100$), thus it becomes apparent from Fig. 1.6b that most of the electrons scattered by the heavy element B are *intercepted* by the objective aperture, or rather their intensity distribution (in Fig. 1.6b) is cut off at the angle α. The few electrons which still pass through the aperture because of smaller scattering angles produce image contours and possible image details in the inclusion B according to both the Abbé theory and the laws of geometric optics. As shown earlier for both theories, it is essential that rays emanate from each sample area in different directions. The *intensity distribution* of the electrons reaching the fluorescent screen is determined exclusively by the number of electrons either passing or being stopped by the objective aperture, depending on the amount of

[4] Because of the analogies in the ray paths, the electromagnetic EM lenses are often schematically sketched in the form of glass lenses.

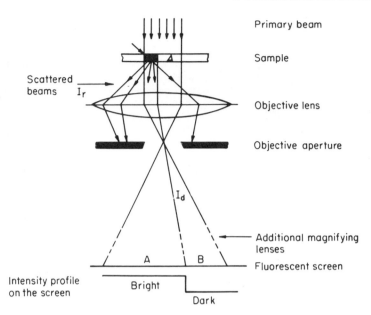

Fig. 1.7. Image formation in amorphous samples (schematically). Real size relations: distance sample − lens ≈ lens − aperture = focal length = 2700 μm; illuminated sample diameter = 1–20 μm, objective aperture diameter = 20–50 μm.

local scattering. The intensity distribution (bottom of Fig. 1.7) therefore *results in relative darkness in the image area from* B, *because very many electrons have been lost from sample area* B *due to the strong scattering (rays intercepted by the aperture).* Here lies the real key to the understanding of image formation. If I_0 is the primary beam intensity, and I_r the intensity lost by scattering, in a given sample area, then the following theorem holds:

The image brightness (of each image detail) is determined by the intensity $I_d = I_0 - I_r$ *of those electrons leaving the lower specimen surface which pass through the objective aperture.*

The qualitative terms "light" and "dark" can be replaced by the "contrast" K of an image which can be quantitatively defined. There are two common definitions:

$$K = \log(I_0/I_d) \tag{1.4a}$$

and

$$K = (I_0 - I_d)/I_0. \tag{1.4b}$$

In both definitions, $K = 0$ for $I_d = I_0$ (i.e., complete brightness or primary beam intensity). On the other hand, for complete darkness, i.e., $I_d = 0$, according to (1.4a) $K = \infty$ and according to (1.4b) $K = 1$.

For an elementary explanation of image formation by scattering so far only the effect of the atomic number (charge number Z) has been considered. Based on the mechanism of Coulomb scattering, however, it is clear that generally the more electrons are lost by scattering,

- the larger the atomic number Z,
- the larger the sample thickness t to be penetrated,
- the more atoms are present for interaction in the specimen film (density ρ; ρt is also called the "mass density"),
- the smaller the aperture angle α (Figs. 1.6b and 1.7), and
- the smaller the energy of the electrons (beam voltage U).

These factors, which affect the amount of scattering, result in an image contrast superficially analogous to the absorption contrast of light microscopy. Therefore it is called *absorption contrast* or *structure factor contrast*. It should be rememberd, however, that true absorption of electrons in the sample practically does not occur (see above), but that it is scattering which produces the effects similar to the absorption of light.

The indicated relationships can be described [3] by the equation

$$I_d/I_0 = e^{-K} = \exp[-c(\alpha, U, Z)\rho t]. \tag{1.5}$$

The function is not a linear function of α, U, and Z. (The relationship can be experimentally determined, for details see Reimer [3, p. 162].)

As an important consequence, these relations show that *contrast* can be *increased in two ways:* first by using a smaller objective aperture, which therefore is also called the *contrast aperture*, and second by using a lower voltage U (the latter in analogy with *x-ray radiography* examinations). In both cases, however, the total image intensity is reduced.

1.6.2. Crystalline Specimens

While for amorphous specimens, the scattering of electrons at the individual atoms could explain the phenomena observed, for a crystalline specimen, the wave nature of the electrons must be considered. According to the Huygens–Fresnel principle, all atoms operate as sources of secondary waves radiating in all spatial directions. (The physical cause is again the scattering process.) Due to the periodic atomic arrangement in the crystal lattice, periodic path differences occur between neighboring scattering centers. These path differences give rise to phase differences. If one considers a given direction in space as an independent variable and asks for the intensity of the secondary waves in this direction, the laws of superposition of wave trains with equal wavelengths, but with different phases, are applicable. (In addition, possibly different amplitudes have to be taken into account, when various kinds of atoms are present.) The precise calculation of the scattered intensity in a given direction will be the subject of Chapter 4 (kinematical theory). At this point, however, a special case, the most important one, must be treated. This is the case in which the phase difference between neighboring wave trains is exactly one wavelength, resulting in the maximum possible constructive inter- ference: namely *Bragg diffraction*. The Bragg equation is as valid for electrons as for x rays and considering its fundamental importance to all of the following, it will be briefly derived here.

Figure 1.8 schematically shows atoms arranged on crystal *lattice planes* with interplanar spacing d. The primary ray (of wavelength λ) impinges at an angle θ on these planes. Constructive interference occurs when a path difference of an integral multiple of the wavelength (i.e., $n\lambda$) exists between the two rays "reflected,"[5] respectively, from the second and first lattice plane. This path difference is $2x$; it is the (twice appearing) distance x in Fig. 1.8. The figure also shows clearly that $\sin \theta = x/d$. With the stipulation that $2x = n\lambda$, the *Bragg equation* is

$$\boxed{n\lambda = 2d \sin \theta}$$ (1.6)

Only when the *angle* θ between lattice planes and incident beam di- rection obeys the *Bragg* equation, does constructive interference occur. The angle θ therefore is also called the *angle of specular reflection*.

The angle between the extension of the incident *primary beam* and the diffracted *secondary beam* is the *diffraction angle*. According to Fig. 1.8, it is 2θ.

For the remainder of this book, calculations will always be made using $n = 1$, signifying the 1st order of the Bragg interference. Higher orders will be taken care of by multiplication by appropriate factors of the Miller indices. It should also be pointed out that, according to colloquial usage which has been accepted though not being completely correct, it is common to refer to the beam as having been "reflected" rather than "diffracted" under the Bragg condition. Actually, the term "reflection" is valid only for single lattice planes.[5]

It is now simple to prove the proposition stated in Section 1.4, that *Bragg reflections from normal metal samples do not pass the commonly used apertures* (up to 50 μm). The smallest possible Bragg angle, for

---

[5] For one (isolated) lattice plane the reflection law is valid, i.e., according to Huygens' principle the maximum intensity occurs when the angle of incidence equals the angle of reflection. In contrast to the Bragg angle θ for three-dimensional lattices, the optical re- flection angle can have arbitrary values.

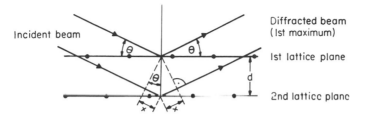

Fig. 1.8. Derivation of the Bragg equation.

example, for diffraction by aluminum occurs for the (111) lattice planes. The interplanar spacing d is determined by the equation for cubic lattices:

$$d = a/(h^2 + k^2 + l^2)^{1/2}, \qquad (1.7)$$

where h, k, and l are the Miller indices [(111) in this case]. Using the lattice constant for Al, $a = 0.405$ nm, one obtains the (111) spacing $d = 0.234$ nm. For 100 kV electrons with a wavelength $\lambda = 0.0037$ nm and $n = 1$, the Bragg equation yields $\sin \theta \approx \theta = 0.008$, $\theta = 0.46°$, or $2\theta = 0.92°$.

In Figure 1.9a, the ray paths in the vicinity of the objective lens are illustrated for the case of a crystalline sample, analogous to Fig. 1.7 for an amorphous sample. The general remarks made for that case apply also here. The primary beam is incident upon the sample from above. The beams diffracted through an angle of 2θ are focused by the objective lens to a *diffraction spot* in the back focal plane. In Fig. 1.1, the diffraction spot was indicated by the focal point F_1, whereas F_0 is the focal point of the undiffracted rays. The sample is located only slightly above the front focal plane. Consequently, the diffracted beam, after passing through the lens, is nearly parallel to the axis. Under these conditions the geometry of Fig. 1.9a (left diagram) shows that the *diffraction spot is intercepted by the aperture diaphragm whenever* $2\theta > \alpha$. (The distance, specimen − lens ≈ lens − aperture = focal length f, and $\tan \alpha$ = aperture radius/f.) Since with $f = 2.8$ mm and an aperture opening of 50 μm, α is only 0.5°, the condition $2\theta > \alpha$ is satisfied for the Al (111) reflection ($2\theta = 0.92°$) and for all higher indexed Al reflections. With Eq. (1.6) one can calculate the "critical" lattice spacing, i.e., that for which $\alpha = 2\theta$. For the 50 μm aperture ($\alpha = 25/2800 = 0.90 \times 10^{-2}$) and with 100 kV electrons ($\lambda = 0.0037$ nm), $d_{crit} = 0.0037/(2 \times 0.45 \times 10^{-2}) = 0.41$ nm.

All the prerequisites for a qualitative explanation of image contrast from crystalline samples have now been given.

To begin with, the three-dimensional, periodic structure formed by the

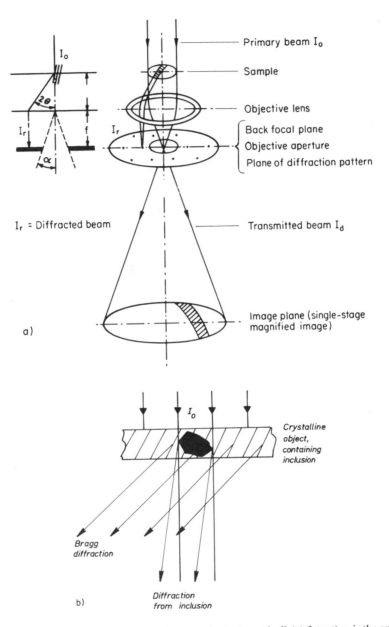

Fig. 1.9. a) Image formation in crystalline samples (schematically) (cf. text). α is the aperture angle and θ the Bragg angle. (See also Fig. 1.1.) b) Diffraction at crystal lattice planes (Bragg diffraction) and diffraction at the contours of inclusions or other object details (Fraunhofer diffraction). There is no basic (physical) difference between both kinds, only a quantitative one. Diffraction angles in the electron microscope are $\lesssim 1°$.

crystal lattice, which causes Bragg diffraction, can be regarded as the special three-dimensional case of Fraunhofer diffraction at a line grating discussed at the beginning of Section 1.2. No fundamental difference exists between the one- and three-dimensional cases. For direct imaging of the crystal lattice planes, at least the first diffraction maximum has to pass through the objective aperture according to the Abbé theory. As calculated above (for 100 kV electrons and a 50 μm objective aperture), this occurs only when the lattice plane spacing $d > 0.4$ nm. The *resolution limit thus calculated for direct imaging of lattice planes* is approximately in accord with the limit of electron microscopic resolution previously discussed [Eq. (1.3), as well as Section 1.3]. Since the *lattice plane spacings of most metals and alloys* are smaller than 0.4 nm, they *cannot* normally, i.e., under the operating conditions discussed, *be resolved*.

In order to lower the resolution limit somewhat further, as, for example, in Fig. 1.5b, some special imaging techniques must be applied. These include extremely accurate lens corrections, somewhat larger objective apertures, and oblique illumination. The latter makes use of the fact that the resolution limit can be lowered from 0.4 nm by a factor of 2 if oblique incidence by the angle θ is used for the primary beam I_0. This follows from the geometry of Fig. 1.9a (left). In this area of high resolution, however, special conditions also exist with reference to the theory of image interpretation which go beyond the scope of this presentation (phase contrast, see, for example, Thon [16]).

All *object details* in crystalline samples whose dimensions are *larger than 0.3–0.4 nm* (for example inclusions or lattice defects) are resolvable in the manner discussed, according to the Abbé theory. The 1st-, 2nd-, or higher-order diffracted beams can pass through the objective aperture. With respect to the *crystal lattice,* these diffracted beams lie in the immediate vicinity of the *zero* order (Bragg) diffraction maximum (Fig. 1.9b).

Although in crystals with lattice spacings $d < 0.4$ nm, Bragg diffraction by itself does not produce an image, it plays a fundamental role in *image contrast,* i.e., in local image intensity because it *removes essential intensity from the primary beam*. This applies to all those sample areas where any lattice plane is oriented more or less correctly for Bragg reflection. *Thus, diffraction in crystalline objects plays a role analogous to scattering in amorphous objects, and everything said there about contrast applies correspondingly here.*

Assume that in the specimen illuminated by the primary beam I_o, as shown in Fig. 1.9a, the cross-hatched region is in the correct Bragg reflection orientation for a given lattice plane. This can be a specific grain in a polycrystalline sample as, for example, grain 1 in Fig. 1.10a. Another frequently occurring case is that of a thin single crystal which is slightly bent. Then only a band-shaped region (as shown in Fig. 1.9a) may be in the correct orientation for a given Bragg reflection (*hkl*). Such dark bands

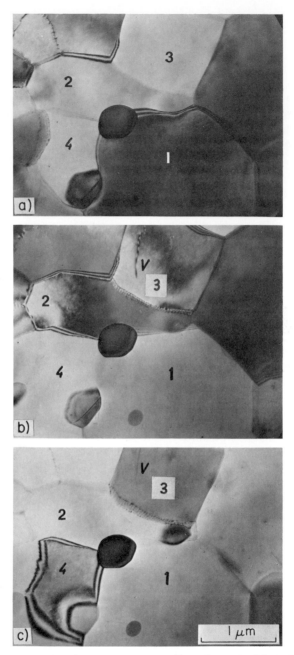

Fig. 1.10. Demonstration of orientation contrast (effect of sample tilting). With regard to (b) the sample is tilted 4° in opposite directions in a) and c) respectively. Alloy: Al–4% Cu. Heat treatment: solution annealed + 90% cold rolled + 9 hr 300°C. The subgrain structure formed contains precipitated θ particles (Al_2Cu), often at grain boundaries. Electrolytically thinned foil (preparation see Chapter 2.1). (Micrograph: K. H. Sieberer, 20,000 × .)

are called "interference fringes" or "extinction contours" (for example, see Fig. 2.10, area S). They will be analyzed in more detail in Chapter 4.

In both cases, so much intensity I_r is removed from the primary beam by the Bragg reflection that the image of the respective sample area appears *dark*. This kind of contrast is called *"diffraction contrast"* or *"orientation contrast."* Due to the very small diffraction angles ($\theta \approx 1°$), a change in orientation of $<1°$ is capable of causing a contrast change from light to dark. Diffraction contrast is therefore very dependent on how precisely a sample area is oriented for Bragg reflection. This is demonstrated in Fig. 1.10. Grain 1, which appears dark in a, brightens after tilting the sample (in this case by about 4°) and appears light in b. Conversely, grains 2 and 3 first appear light and without contrast, while after the tilting, in b they appear shaded, or "in contrast" according to the local lattice orientation. A dislocation V in grain 3, which is not visible in a, is clearly visible and in strong contrast after tilting (b). On the other hand, grain 4, in *both* a and b, is light and completely without contrast. This grain attains good contrast only in c by tilting about 4° in the opposite direction.

Note also the *"fringe contrast" of the grain boundaries (thickness interference fringes)* which will be explained in Section 4.5. These grain boundary fringes, too, can be made to appear or disappear by tilting. This is especially striking for the grain boundaries of grain 4. Figures 1.10(a)–(c) are therefore an instructive example of how important it is in practical microscopy to tilt the sample in order not to overlook important details, such as grain boundaries, dislocations, etc. Usually very small tilt angles, considerably smaller than 1°, are sufficient for these effects.

In all of these cases, the *rays deflected* by Bragg diffraction are focused to a *spot* in the back focal plane by the objective lens (coil), indicated as a ring in Fig. 1.9a. This is the (hkl) diffraction spot. With appropriate orientation, i.e., with a low index crystallographic direction approximately parallel to the primary beam (more in Chapter 3), additional diffraction spots appear in the back focal plane. These are shown in Fig. 1.9a; they form the complete *diffraction pattern*. This is a spot pattern when the illuminated foil is a single crystal, and it is a ring pattern (Debye–Scherrer rings) when many crystallites are illuminated by the beam.

As previously discussed, the main beam (zero maximum with reference to Bragg diffraction) passing through the aperture creates the enlarged *image of the sample* in the image plane of the objective lens. This situation is shown in Fig. 1.9a. The intermediate and projector lenses further enlarge the image, as discussed in Section 1.4 and shown in Fig. 1.3a. The final image thus produced from crystalline samples on the screen is the *bright-field image*. A *dark-field image,* on the other hand, is produced when the contours of an object (Fraunhofer diffraction) are imaged not by the zero Bragg diffraction maximum, but instead by any nonzero Bragg reflection ("image formation due to a defined Bragg reflection").[6] This is achieved, for example, by translating the objective aperture so that, instead of the primary beam, the desired diffracted beam passes through the aperture (Figs. 1.11b and 1.12). Light and dark reverse their contrast, but only those regions are bright which are oriented correctly for diffraction.

Figure 1.11 is a simplified diagram of the beam paths between the sample and the objective aperture. Below the objective aperture, the beams continue as shown in Fig. 1.3a for bright field image formation. By displacing the aperture diaphragm, however, the image quality is degraded because off-axis beams are used for imaging (spherical aberration); a better, although somewhat more laborious, procedure is to tilt the beam (electron gun plus condensers), which is possible with most microscopes (Fig. 1.11c). Because the imaging beam is axial, the image quality is preserved.

With the newer types of EM the necessary oblique incidence of the primary beam can be achieved easily and precisely with an additional pair of built-in deflection coils. The adjustable field of these coils "bends" the primary beam twice so that the diffracted beam desired for dark field imaging is aligned precisely along the microscope axis. If the deflection system is switched off, the bright field image reappears with the axial alignment previously set for the primary beam.

Switching from enlarged image to diffraction pattern: In order to obtain the entire diffraction pattern on the final image screen, it is merely necessary to completely remove the objective aperture and to reduce the excitation of the intermediate lens. In this way, the focused plane is not that of the single-stage enlarged image, but the plane of the diffraction pattern. The beam path produced in this manner is shown in Fig. 1.3b. In addition, one can select *small regions* of the sample containing interesting inclusions or similar details by inserting the *selector aperture* in order to obtain diffraction from only these areas ("selected area diffractions," abbreviated "SAD").

The operation of the selector aperture can be understood with the help of the beam diagrams in Fig. 1.3 or Fig. 1.9a. The cone angle of the rays passing through this aperture is directly proportional to the diameter of the area illuminated in the sample. The proportion-

---

[6] The Fraunhofer diffraction from all other specimen details (e.g. contours of particles) is contained as "information" in the vicinity of this diffracted beam, as well as in the primary beam.

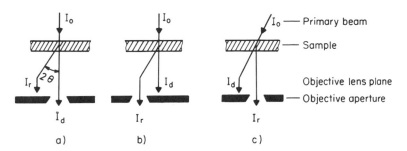

Fig. 1.11. Bright and dark field imaging. Simplified ray path (focusing disregarded). (a) bright field imaging, (b) dark field imaging by displacing the objective aperture, and (c) dark field imaging by tilting electron gun or primary beam, respectively.

ality factor is the magnification factor of the objective lens with reference to *the single-stage enlarged image in the plane of the selector aperture*. This factor is about 20. It should not be confused with the magnification of about 200 mentioned in Section 1.4, which is obtained when the sample is imaged on *the final screen* by the objective lens alone (all other lenses being turned off).

A selector aperture of, for example, 20 μm in diameter will enclose, therefore, a circular area of only 1.0 μm in diameter in the plane of the object. This area alone will contribute to diffraction. The possibility to identify such small areas or amounts of substances and to ascertain structural differences in this size scale is one of the specific strengths of the transmission microscope for many problems in materials science and metal physics as well as for other applications with transparent, crystalline samples. Furthermore, the possibility of immediately switching from the diffraction pattern to the image itself of the same sample area is very useful.

An example of bright- and dark-field images of the same sample area of an Al–Cu–Mg–Pb alloy is shown in Fig. 1.13. The alloy contains plate-

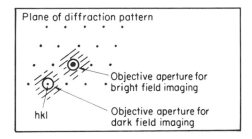

Fig. 1.12. Bright and dark field imaging: position of objective aperture in plane of diffraction pattern. The primary beam is denoted by the large black dot.

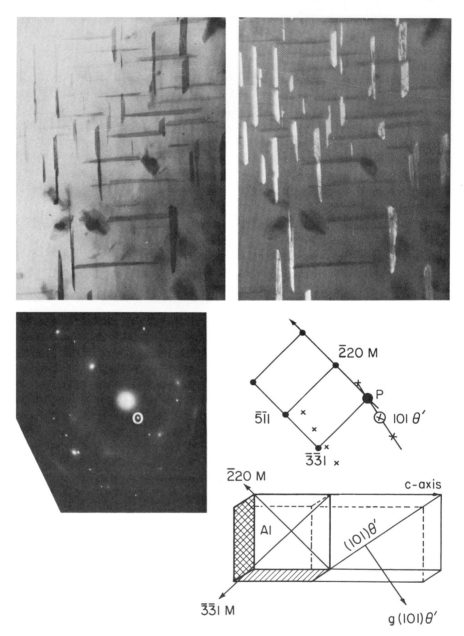

Fig. 1.13. Bright field image (left), dark field image, diffraction pattern, and position of the unit cells of matrix (aluminum) and plate-shaped precipitates (θ' phase), 20,000×. Material: AlCuMgPb alloy aged 2 hr 300°C. In the schematic diffraction pattern the reflections of the matrix (*M*) are shown as points, the precipitate reflections as crosses. *P* denotes the primary beam.

like precipitates of the θ' phase (Al_2Cu), which are situated on the {100} planes of the Al matrix. The diffraction pattern of this sample area is shown on the left and is drawn schematically on the right. The pattern was analyzed, including consideration of the magnetic rotation, using the methods presented in Chapter 3. The result is that the [116] direction of the Al lattice is parallel to the primary beam I_0. This slightly tilted [001] orientation of the cubic unit cell of aluminum is illustrated at the bottom of Fig. 1.13. The θ'-plates lying on the two hatched, or shaded, cube planes are clearly visible in the bright field image. (If [001] were $\| I_0$, then the plates would appear only as a thin line.)

The θ'-phase has a tetragonal structure with a = 0.405 nm and c = 0.58 nm. The c axis always is normal to the plane of the plate. The dark field picture was produced with the circled θ'-reflection which is also circled in the schematic diffraction diagram. Those (101) planes of the corresponding unit cell which generated this reflection are shown in the sketch in the position corresponding to the pictures above. As can be seen, this is the particular one of the three possible positions of the θ'-cells whose c-axis points to the right. And in fact, only these (hatched) plates normal to that axis are illuminated *brightly* in the dark field image of the circled (101) θ'-reflection. The other precipitates (shaded) remain dark.

It is one of the advantages of *dark field pictures* that one can differentiate certain crystal types or orientations from the remainder of the crystals, even when they are not so clearly distinguishable in image appearance, as in the case of Fig. 1.13.

Finally, with reference to the image contrast of crystalline samples, it should be noted that the effects of mass contrast, described in Section 1.6.1, obviously can be superimposed on the effects discussed in this chapter.

1.7. High Voltage Electron Microscopy

The first generation EMs, 1932 until about 1950 (Section 1.1), operated with acceleration voltages of approximately 40–70 kV. Later, the maximum voltage was 100 kV for commercially produced EMs. This was a voltage which could still be produced economically and which could be sufficiently stabilized. In such microscopes, samples approximately 50–300-nm thick, depending on their atomic weight, were electron transparent. Irrespective of the sometimes difficult preparation of such specimens, the fundamental drawback of such thin foils is that often they cannot be regarded as representative of bulk material. For example, some lattice defects, such as dislocations, are altered (via image forces) or disappear altogether because of the proximity of the surface. Therefore, in order to penetrate thicker samples, higher acceleration voltages are desirable.

Since the beginning of the 1960s, therefore, a "third generation" of EMs has made its appearance. Operationally safe and sufficiently stabilized high-voltage generators had been developed for the emerging nuclear technology. These generators (Cockcroft–Walton or cascade generators) could also be used for commercially produced EMs. The most important properties of these high-voltage electron microscopes (200–1000 kV) are presented briefly. The *advantages* are

a) *Thicker specimens* are electron transparent: The transparency of inorganic crystalline materials is a function of both the accelerating voltage of the EM and the atomic number of the material. For example, when comparing 100 kV and 1000 kV, the thickness increases by a factor of 3 for steel and 8 for silicon (K. Urban [18]).

The increased transparency is based on the reduced effective cross section for elastic and inelastic scattering. For amorphous material, the reduced "absorption" (compare Section 1.6, contrast formation) can be described by a law of the type $I_d = I_0 \exp(-\epsilon t)$, where t is the sample thickness and ϵ the "effective absorption coefficient." However, as was emphasized in Section 1.6, this is not true absorption, but mostly scattering. At present, it appears to be certain that ϵ is proportional to $(c/v)^2$ for medium–high voltages. Since v approaches c asymptotically, the improvement attainable with increasing high voltage becomes relatively smaller.

b) Corresponding to the shorter wavelengths (Table 1.1), *resolution* increases. Theoretically, it should, for example, be 0.09 nm at 1000 kV, on the basis of Eq. (1.3) with $A = 0.43$ and the assumptions as in Section 1.5. However, in practice, this increase of resolution cannot be fully realized, because sufficient stabilization of such high voltages still presents a problem (chromatic aberration).

c) As a result of the higher electron energies, the *energy loss* of the electrons passing through the sample is diminished as a consequence of the smaller (real) *absorption*. *Less radiation* is absorbed and *less heating* of the sample occurs with 1000 than with 100 kV (compare Section 1.6.1).

In normal EM operation, the sample surface is contaminated due to the deposition of hydrocarbon molecules (pump oil vapor) cracked by the electron beam. For "harder" electron beams, i.e., high-voltage electrons, such contamination is reduced, because there is a weaker mutual reaction between these electrons and the residual gas molecules. For 100 kV, the contamination layer on the specimen has a growth rate of a few tenths nm/s. Contamination can be greatly reduced by the use of a cooling stage (commercially available) cooled by liquid nitrogen.

In contrast to these *advantages* there are, however, some *disadvantages*.

a) In order to keep the instruments from becoming too large, considerably stronger lens fields are necessary (about 20 kG) if the focal length, for example, of the objective lens is to be held to approximately 5–6 mm. Because of these larger lenses as well as the larger high voltage generators, the *dimensions and cost* of such microscopes increase, the dimensions approximately proportional to the square root of the high voltage, the cost directly proportional! For high voltages ≥500 kV, specially constructed buildings are necessary.

b) The *image contrast* diminishes with increasing high voltage.[7]

c) The *x rays* produced at the final image screen are so hard, that very thick lead glass shielding is necessary for radiation protection. In some installations, television transmission of the final image or electronic intensification is used. In either case, the beam current in the microscope can be considerably reduced.

d) Above a certain displacement threshold energy U_d, which depends on the substance irradiated, the specimen—regardless of the reduced overall radiation absorption [see advantages, point c)]—can experience *specific irradiation damage* caused by the high-energy electrons. Examples are ionization in insulators as well as Frenkel defects. (Formation of an interstitial and a vacancy by electron impact.)

Makin [17] observed in aluminum in a 500 kV, 0.1–0.4 μA electron beam, focused to 5 μm diameter in the sample, that numerous dislocation loops (vacancy and interstitial loops) were formed by Frenkel defects after 5 min and up to 1 hr. On the other hand, this effect offers the interesting possibility to accurately measure the threshold energy for the formation of Frenkel defects.

In summary, it can be said that high-voltage electron microscopes will certainly become increasingly more important. The range of 200–500 kV should prove to be especially interesting, since it can be theoretically shown that the most important advantages [points a) and c)] improve only slowly above 500 kV, so that the cost–benefit ratio above 500 kV becomes increasingly less favorable (The present success of high-voltage electron microscopes as tools for the in situ production and observation of irradiation damage in materials has, nevertheless, made it highly desirable to go to accelerating voltages of ≧1000 kV).

For further information, the interested reader is referred to the more recent specialized literature [18].

---

[7] This follows theoretically from Chapter 1.6 (elastic scattering). Practically, however, the image contrast is often *increased* because of the decreased inelastic scattering, see Warlimont [18].

1.8. Surface Electron Microscopes

There are various possibilities to obtain enlarged images of surfaces of bulk, nontransparent samples with the help of electron beams. These are quite diverse in principle and significance:

a) In the *reflection electron microscope,* the surface is irradiated with an electron beam at an angle of only a few degrees, i.e., very obliquely (Fig. 1.14a). These electrons are *scattered* predominantly in the direction symmetric to the sample normal and can then be used, in a beam path analogous to the one in the transmission microscope, to enlarge the image. The grazing incidence is necessary for reasons of intensity, since the scattered intensity diminishes rapidly with larger angles.

This method has two considerable disadvantages: First, the attainable resolution of only 20–50 nm is very poor in comparison with the transmission microscope and, second, the sample surface is imaged with severe distortion because of the very oblique "observation direction." Because of these reasons, this type of microscope plays only a minor role.

b) The *emission-electron microscope* avoids the disadvantage of the very oblique projection since it utilizes for image formation electrons that are directly emitted from the sample surface and leave the latter almost normally, Fig. 1.14b. The subsequent beam path (acceleration by the anode, enlargement by the objective lens, projector lens, etc.) again corresponds to the previously described electron microscope. The emitted electrons are called *secondary electrons;* their emission is caused by *primary particles* which are incident obliquely from the side. The following particles can be used as primaries:

- *Ions*
- *Electrons*
- *Photons* (Emission of electrons from metals by light of short wavelength is called the *photoelectric effect, photoemission.*)

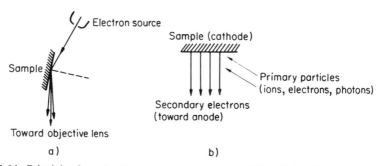

Fig. 1.14. Principle of a) reflection electron microscope and b) emission electron microscope.

• *Thermionic emission.* The natural self-emission of electrons from the surface at sufficiently high temperatures is termed thermionic emission (self-emission). The emissive work function can be decreased by the well known method of barium deposition (by evaporation).

From among these possibilities, the emission due to photon excitation by *ultraviolet (UV) light* has gained in importance in recent years. Up to four intense UV lamps are used simultaneously to illuminate the sample. Due to technical progress it has become possible with this type of microscope to obtain a resolution of 15 nm. Although this is still over ten times poorer than that of the transmission microscope, the emission microscope has the following principal advantages [19]:

Without special preparation, metal surfaces can be magnified more highly (up to 12,000×) than in the light microscope while retaining the familiar appearance of optical micrographs. In addition, it is possible with an ion etching procedure in the emission EM to repolish and etch sample surfaces which have become rough during the course of examination. The main importance of the UV microscope, however, may be its application to observations in the *high-temperature region* (up to the melting point of the sample for long term experiments). The method appears to be predestined for this application because of the thermionic emission occurring at high temperatures. Furthermore, it is possible to record dynamic reactions *cinematographically* with a vacuum film camera. Finally, by the deliberate introduction of chemically active gases, the reactions of metal surfaces with these gases can be observed [19]. Figures 1.15a and b give examples of pictures taken with an emission electron microscope.

Three different types of contrast can be distinguished [19], attributable to different causes. *Orientation contrast* arises from the different angles between the crystallite axes and the optical axis of the microscope, *material contrast* results from the different work functions of the various phases, and *topographical contrast* arises from the local inclination of surface elements to the optical axis of the microscope.

c) The *scanning electron microscope* (SEM) is not an EM in the sense of the previously discussed instruments, since it does not produce enlarged images by means of electromagnetic lenses in analogy with light optics. Yet it is called an "electron microscope" because it produces a highly magnified image by means of electrons.

The principle is shown in Fig. 1.16. First, an electron beam is produced by an electron gun just as with the normal EM and is focused to a point on the sample surface by two condenser lenses. Between the two condenser lenses, a "stigmator" is placed to correct the astigmatism. The second condenser lens is occasionally also termed—although not completely correctly—the "objective lens." The essential point is that it is

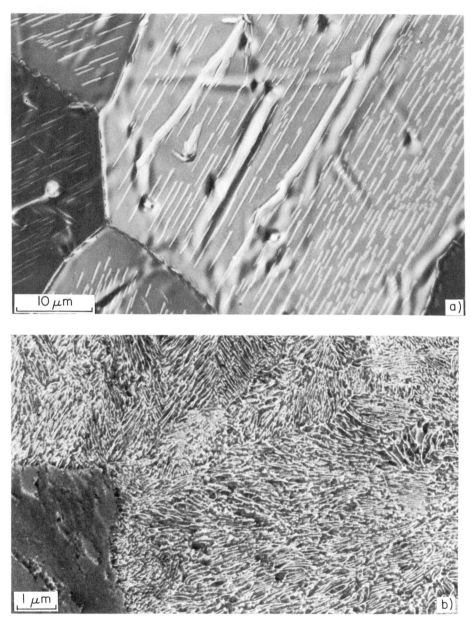

Fig. 1.15. Micrographs made with an emission electron microscope. a) molybdenum with carbide lamellas and scars of former carbide growths. Electrons generated by thermal emission ($\approx$1600°C). $C_2 H_2$ atmosphere during observation. 2000×. b) Steel C110 (1.1%C), quenched in oil from 1100°C. Pearlite (fine lamellas), martensite in lower left corner (electrolytically polished, *not* etched). Electrons generated by ultraviolet light at room temperature. 10,000×. (Micrographs by Fa. Balzers AG., taken with a METIOSKOP KE 3.)

34

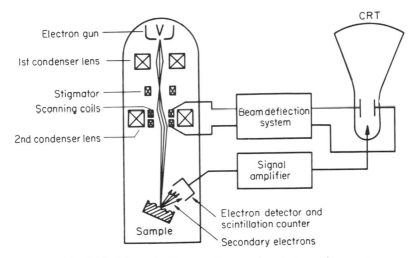

Fig. 1.16. Schematic diagram of a scanning electron microscope.

possible to focus the beam to an *extraordinarily small diameter of only 10–20 nm* (compared with approximately 1000 nm = 1 μm for the electron microprobe and approximately 5 μm for an ordinary transmission microscope). In the second condenser lens (in some designs below the condenser), two pairs of *scanning coils* are present. These deflect the focused beam so that it *scans* a square region of the sample surface, similar to the scanning of the image in a television camera. The focused beam produces *secondary electrons* on that sample spot on which it impinges at any moment. These electrons are collected and counted by an electron detector which is located above and to one side of the sample. The detector consists of a grid with accelerating (bias) potential in front of a scintillation counter which transforms every electron into a light flash. A light guide conducts the flashes to a secondary electron multiplier which transforms them into electrical pulses. These are amplified by a signal amplifier and used to modulate the *brightness of a cathode ray tube* (CRT) by means of its control grid. The deflection system for the electron beam in this CRT is controlled by the same generator which also controls the deflection coils in the scanning microscope column.

Thus the principle is clear: Every point on the object (sample) is transposed to a point on the CRT, just as with television. The brightness, i.e., the intensity of the secondary electrons emitted by the spot in question and registered by the detector, depends critically on the local surface relief because of the asymmetric location of the detector. Thus, small holes are "imaged" black because electrons emitted from the inner surface of the holes do not reach the detector. Areas facing the detector appear brightest, while "shadows" with lesser intensity are produced behind protrusions, ridges, and similar features. Secondary electron emission also depends on the angle between primary beam and sample surface. The geometrical relief of the sample, therefore, is magnified into a relief map or topographic image on the CRT (topography). The linear *magnification* in the scanning microscope is given by the ratio of the scanned length in the CRT (for example, 100×100 mm) and the length scanned on the sample (minimum approximately 5×5 μm). Thus, *useful magnifications* (continuously variable) are from $50\times$ up to approximately $20,000\times$.

Although this principle of the scanning microscope has been known for 30 years, its practical development into commercial instruments has been made possible only after the above mentioned focusing of the electron beam to a spot size of only 15 nm or even less became technically feasible. The *resolution* is essentially identical with the spot size, although somewhat poorer because of the scattering processes in the sample. The *accelerating voltage* for the electron beam is variable between 5 and 50 kV. The resolution diminishes with decreasing voltage.

Since the focused beam is so narrow, effective focus is maintained over a certain depth range. Consequently, the resulting images have an *extraordinary depth of field*. This is, for example, 35 μm at $1000\times$ magnification, much larger than in the light microscope. This additional advantage, besides the three-dimensional appearance of the pictures, for which Figs. 1.17 and 1.18 are given as examples, contributed to the rapid distribution of this new type of microscope.

The samples, if they are conductors, can be investigated, without any sort of preparation, in their original condition. Only nonconducting samples, similar to microprobe samples, must be coated with a thin layer of a conducting material (for example, gold or carbon).

It should be mentioned here that, in addition to the secondary electrons with maximum energies of 50 eV, backscattered electrons with approximately the same energy as the primary electrons are also emitted from the sample surface. They too can be used for "imaging" the relief (topography). The resolution is worse than with secondary electrons because the backscattered electrons are emitted from a deeper scattering cone in the sample. However, a certain differentiation of the elements (composition) in the respective sample spot can be obtained, since the number of high-energy, backscattered electrons increases with the atomic number of the element in the surface. The majority of pictures with the scanning microscope are made with secondary electrons because of the better resolution.

Finally, the scanning electron microscope offers still further possibilities which can only briefly be mentioned here. Instead of secondary or backscattered electrons, other signals can be used to modulate the local brightness of the image screen. These include:

• the electromotive force which the focused electron beam produces, e.g., at the collector of a semiconductor-transistor;

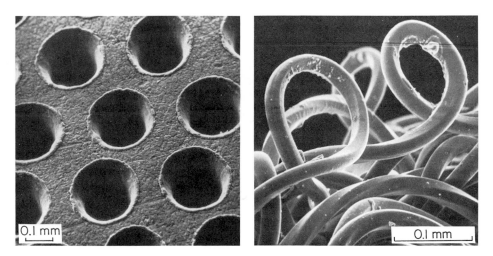

Fig. 1.17. Micrographs made with a scanning electron microscope: a) holes in Nimonic, "drilled" with an electron beam (Samples: K. H. Steigerwald, 80×.) b) Nylon fibers (sock), 210×. (Both micrographs: courtesy H. J. Dudek, Physikalisches Laboratorium Mosbach, Germany, taken with a Stereoscan-EM, Cambridge Ltd.)

- the transmitted electrons, if the sample is sufficiently thin (bright-field image by means of primary beam, dark-field image by means of diffracted electrons);
- the absorbed electrons (current image);
- the characteristic x-rays (mapping of elemental distribution).

For further details, see the literature of the manufacturers and Reimer [11].

1.9. Scanning Transmission Electron Microscopy (STEM) and Energy-Dispersive X-Ray Analysis (Elemental Microanalysis)

Scanning transmission electron microscopy (STEM) is an important additional application of a scanning technique within a conventional transmission electron microscope. This mode of operation is possible, e.g., by using a so-called "single field condenser/objective lens" (Riecke–Ruska lens) consisting of two lens parts excited by a single coil (the objective coil), as in Fig. 1.19. The upper lens is used to focus the electron beam onto the sample in a spot of only 3 nm diameter (or less). Since the spot size determines the spatial resolution in scanning electron microscopy, a sufficiently small spot is a prerequisite for the success of STEM. The more recent highly sophisticated developments in electronics have been responsible for achieving such extremely small spot sizes. Whereas in TEM a parallel electron beam passes through the sample, in STEM the

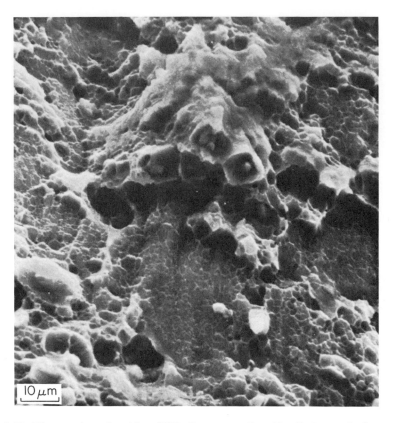

Fig. 1.18. Micrograph made with a SEM: Fracture surface (ductile fracture) of a tensile sample of an Al alloy AlZnMg1. 25 kV, takeoff angle 45°, distance sample–detector 12 mm. The fracture surface looks fibrous, typical for a ductile fracture [20]. Many holes have been formed due to internal necking. These holes almost always form at inclusions giving the impression of a bell, with a "clapper" in it. (Courtesy: Fa. Kontron Gmbh., München, Germany, JEOL-SEM.)

beam is slightly convergent. A pair of deflection coils in the upper part of the microscope operates as the scanner. After penetrating the sample, the beam is slightly divergent; it is then made parallel by the lower lens of the "single field condenser/objective lens", as in Fig. 1.19. An electron detector at the position of the fluorescent screen provides further signal processing analogous to that in the SEM. Since in STEM inelastic scattering of electrons in the sample is of practically no concern (no imaging lens is needed), samples much thicker than with TEM can be imaged. The resolution (the spot size), of course, is about an order of magnitude

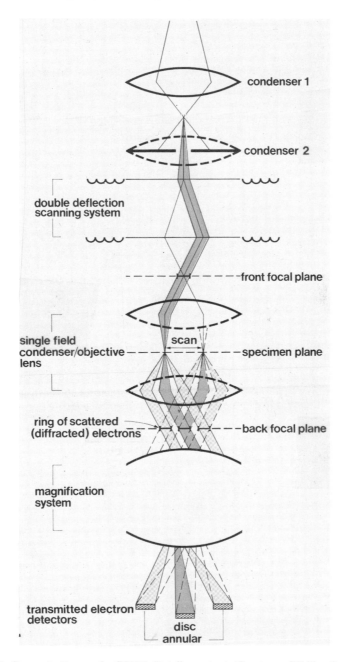

Fig. 1.19. Ray path diagram for STEM. Details see text. (Courtesy of Philips Comp., Eindhoven, Netherlands.)

worse than for TEM. In addition, the so-called top–bottom effect and the beam convergence reduce practical resolution. However, technical improvements to reduce the spot size are still being made.

If the sample is crystalline, Bragg diffraction occurs in STEM as it does in TEM. The undiffracted primary beam (dark gray in Fig. 1.19) produces the bright field image. Diffracted beams (light gray in Fig. 1.19) are focused off-center by the lower part of the objective lens. In advanced systems, an annular detector surrounding the central disk detector is summing up these diffracted beams. As in TEM, dark field images can be obtained in STEM by using the diffracted beams. If the whole annular detector is used, "total dark field" images are produced by the summation of all reflections on a Debye–Scherrer ring. (Presupposition is a stationary diffraction pattern in the back focal plane of the objective lens also in STEM mode.) If a dark field image in the light of a single defined Bragg reflection is desired, the disk detector is shifted to the position of that reflection.

Most modern high-resolution TEM microscopes can be equipped with an optional STEM attachment. Simple switching arrangements change the microscope from the TEM to the STEM mode of operation. In the TEM mode the lower part of the "single field condenser/objective lens" serves as the high-resolution objective lens as discussed earlier (Sections 1.3 and 1.6).

Originally it was thought that the main advantage of STEM was the greater penetrating power of, say, 100 kV electrons, although at the expense of some resolution. Later it was discovered that even better opportunities for gaining additional information about the sample are provided by other operational modes (Fig. 1.20). These include:

Above the sample:

• secondary electron images (as in conventional SEM),
• backscattered electron images,
• x-ray elemental analysis.

Below the sample (in addition to STEM itself):

• electron energy loss analysis.

It is especially x-ray elemental analysis which, in recent years, has become an important and powerful analytical tool in combination with TEM and STEM. In principle, point analysis, area analysis, line scan, and x-ray mapping are possible, analogous to the operational modes of the well-known electron microprobes. The most obvious advantages are:

(i) direct comparison with the corresponding TEM image at magnifications much higher than are possible with a microprobe;

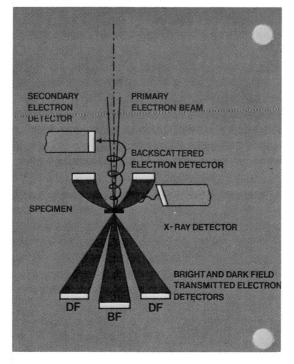

Fig. 1.20. Different types of information, which may be utilized above or below the specimen in addition to normal imaging. (Courtesy of Philips Comp. Eindhoven, Netherlands.)

> (ii) the much higher spatial resolution of the analysis—aptly called "microanalysis" because characteristic x rays (K_α, K_β, L_α, etc.) are excited only from the extremely small sample volume illuminated by the 3-nm-diameter spot (see Fig. 1.22); and
>
> (iii) the possibility of taking special electron diffraction patterns which can give crystallographic information on particles, precipitates, etc., as small as the beam.

In contrast to normal selected area diffraction (SAD) where the area is defined by apertures (see Section 1.6.2, Figs. 1.11, and 1.12), the area in the above so-called *"microarea diffraction"* (MAD) is defined by the narrow beam itself.

The x-ray signal intensity is much lower than in conventional microprobe analysis, even if the beam is spread to a diameter of, say, 5, 10, or 15 nm (sacrificing some spatial resolution). Therefore energy dispersive analysis is preferred over wavelength dispersive analysis (with crystal spectrometer). In energy dispersive (x-ray) analysis (EDA or EDX, also called EDS for energy dispersive spectroscopy) a Li-drifted silicon crystal (covered by a thin, e.g., 7 μm beryllium foil) is used to analyze the spectrum. The x-ray quanta K_α, K_β, etc., having different wavelengths and correspondingly different energies, are converted into electrical pulses with the pulse heights proportional to the respective x-ray energies. After preamplification and amplification, the pulses are classified in a multichannel analyzer and can be displayed, e.g., on a CRT, printer, or plotter. The x-ray energy, which is characteristic for each element, is plotted along the abscissa, the x-ray intensity (an uncorrected measure for the abundance of the element) along the ordinate. For details see the literature on x-ray microprobe analysis [25, 27] and on scanning electron microscopy [11].

To obtain good results with the combination STEM and EDS technique, a number of special conditions must be met. Some of these have to be taken care of by the microscope manufacturer, others by the operator. Because of the high local intensity concentrated in the narrow electron beam, contamination rates due to residual gas pressure are high. Therefore the microscope *vacuum* should be at least as low as 10^{-7} Torr, and the residual fraction of hydrocarbon molecules should be as low as possible. This can be achieved not only with special pumps, but metal seals instead of rubber O-rings, etc. Since an x-ray point analysis with a 3–5 nm beam can take up to several minutes, specimen *drift* has to be negligible. Otherwise, the beam might move off the point of interest during the analysis.

Useful improvements in microscope design are the LaB_6 cathodes (see p. 7) and field emission sources. With their high brightness, the counting (i.e., analysis) time in EDS is significantly shortened for very small particles. The field emission cathodes, however, are quite expensive and have the additional disadvantage of providing lower specific intensity (brightness) in normal TEM illumination, i.e., with the beam diameter spread to 1–10 μm.

An example for a STEM bright field image and correlated EDS spectra is given in Fig. 1.21. The investigation was undertaken to clarify whether crystallization of the metallic glass phase occurs with or without change in elemental composition. Comparison of the spectra, Figs. 1.21b and c, shows that the crystal phase has a lower P and Cr content and a higher Fe and Ni content than the glassy matrix. (For details see von Heimendahl and Oppolzer [26].) These two spectra are an example of a "semiquantitative" analysis, corrected only for background. In a fully quantitative analysis, additional correction procedures have to be carried out to convert the raw x-ray intensities to true chemical compositions. This includes (for details see [28–30])

• material-specific *"ZAF"* corrections (atomic number Z, absorption A, and fluorescence F) and
• several instrument and detector-specific corrections (peak overlap,

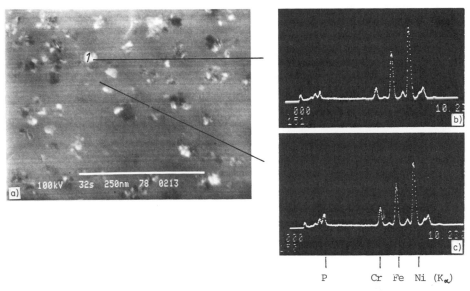

Fig. 1.21. Amorphous metal $Fe_{32}Ni_{36}Cr_{14}P_{12}B_6$ ("Metglas 2826A"). a) STEM dark field image using annular detector, 100 kV beam voltage. Energy dispersive x-ray spectra of b) crystal 1, and c) nearby glassy matrix. Beam diameter 5 nm, beam current 2×10^{-8} A, counting time 150 s. For details see von Heimendahl and Oppolzer [26]. (Taken with a Siemens Elmiskop ST 100 F.)

electronically originated peaks as "summation peaks" and "escape peaks," and the energy dependence of the detector system.

The latter concerns the fact that the relative sensitivity of the whole detector system is not constant for all elements of the periodic system. The relative mass fractions of the two elements, c_1 and c_2, are correlated with the measured (and corrected, see above) intensities I_1 and I_2 by

$$c_1/c_2 = k_{1,2} (I_1/I_2).$$

Examples for k factors are given, e.g., by Cliff and Lorimer [31].

Two other examples for the usefulness of microanalysis with STEM are given in [24a], for commercial steel, and in [24b and c], for a creep-resistant age-hardening aluminum alloy. Von Heimendahl *et al.* [24b] have shown by microanalysis that in the Al alloy 2219 the precipitated θ' phase is enriched with manganese, thus stabilizing the alloy against particle coarsening (Ostwald ripening).

A comprehensive treatment of quantitative EDS is found in the monograph by Chandler [25]. It should be emphasized, however, that in EDS combined with STEM correction procedures are simpler than in the more common applications of microprobe analysis and SEM. In bulk samples the "scattering volume" is pear-shaped and has a diameter of approxi-

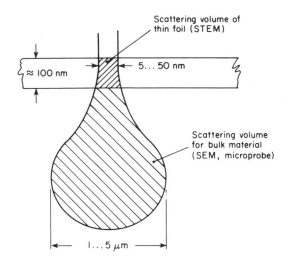

Fig. 1.22. Scattering volumes contributing to the EDS signal, schematically.

mately 1 μm or more (Fig. 1.22). In such a large volume the above mentioned *ZAF* influences affect the characteristic x-ray yield and have to be accounted for. In STEM applications, on the other hand, the sample is so thin (50–100 nm) that only the narrow neck of the scattering volume (Fig. 1.22) is contained in the sample and contributes to x-ray emission. Therefore, the rather complicated and time consuming *Z* and *A* corrections are not necessary in the case of STEM.

References

Books

a) General

1. P. B. Hirsch, A. Howie, R. B. Nicholson, D. W. Pashley, and M. J. Whelan, "Electron Microscopy of Thin Crystals." Butterworths, London, 1971.
2. G. Thomas and M. Goringe, "Transmission Electron Microscopy of Materials." Wiley, New York, 1979.
3. L. Reimer, "Elektronenmikroskopische Untersuchungs-und Präparationsmethoden," 2nd ed., Springer, Berlin and New York, 1967.
4. G. Thomas and J. Washburn (eds.), "Electron Microscopy and Strength of Crystals." Wiley, New York, 1963.
5. R. D. Heidenreich, "Fundamentals of Transmission Electron Microscopy." Wiley (Interscience), New York, 1964.
6. E. Hornbogen, "Durchstrahlungselektronenmikroskopie fester Stoffe." Chemie-Verlag, Weinheim, 1971.
7. L. E. Murr, "Electron Optical Applications in Materials Science." McGraw Hill, New York, 1970.

8. P. J. Goodhew, "Electron Microscopy and Analysis." Wykeham Publ. London, 1975.
9. M. H. Loretto and R. E. Smallman, "Defect Analysis in Electron Microscopy," Chapman and Hall, London, 1975.
10. B. K. Vainshtein, "Structure Analysis by Electron Diffraction" (translated from Russian). Pergamon, Oxford, 1964.
11. L. Reimer and G. Pfefferkorn, "Raster-Elektronenmikroskopie," 2nd ed., Springer, Berlin and New York, 1977.
12. K. W. Andrews, D. J. Dyson, and S. R. Keown, "Interpretation of Electron Diffraction Patterns." Hilger and Watts, London, 1967.
13. G. Schimmel, "Elektronenmikroskopische Methodik." Springer, Berlin and New York, 1969.
13a. P. J. Grundy and G. A. Jones, "Electron Microscopy in the Study of Materials." Edward Arnold Publ., 1976.

b) Especially for Specimen Preparation

14. G. Schimmel and W. Vogell, "Methodensammlung der Elektronenmikroskopie," Ringbuch Hrsg. Wissenschaftl. Verlagsgesellschaft mbH, Stuttgart, 1971.
15. D. H. Kay (ed.), "Techniques for Electron Microscopy," 2nd ed., Blackwell, Oxford, 1965.
15a. I. S. Brammar and M. A. P. Dewey, "Specimen Preparation for Electron Metallography." Blackwell, Oxford, 1966.
15b. P. J. Goodhew, "Specimen Preparation In Materials Science." North-Holland Publ., Amsterdam, and American Elsevier, New York, 1973.

c) Original publications[8]

16. F. Thon, *Physikal. Blätter* **23**, 450 (1967).
17. M. J. Makin, *Phil. Mag.* **18**, 637 (1968).
18. Some basic literature for high voltage electron microscopy:
V. E. Cosslett, Recent progress in high-voltage electron microscopy, *in* "Diffraction and Imaging Techniques in Material Science" (S. Amelinckx, R. Gevers, and J. van Landuyt, eds.), pp. 511 to 549. North Holland Publ., Amsterdam, 1978.
P. R. Swann, C. J. Humphreys, and M. J. Goringe, eds., "High-Voltage Electron Microscopy." Academic Press, London and New York, 1974.
H. Warlimont, *Pract. Metallogr.* **7**, 654 (1970).
K. Urban,. *Umschau Wissensch. Techn.* **78**, 363 (1978).
19. L. Wegmann, *Prakt. Metallogr.* **5**, 241 (1968).
20. E. Dick, *Z. Metall.* **60**, 214 (1969).
21. R. Sinclair, K. Schneider, and G. Thomas, *Acta Metall.* **23**, 873 (1975).
22. R. Sinclair and G. Thomas, *Metall. Trans.* **A9**, 373 (1978).
23. V. A. Phillips, *Acta Metall.* **23**, 751 (1975).
24a. P. Doig, D. Lonsdale, and P. E. J. Flewitt, *Micron* **10**, 277 (1979).
24b. M. von Heimendahl and V. Willig, *Scripta Met.* **11**, 875 (1977).
24c. V. Willig and M. von Heimendahl, *Z. Metall.* **70**, 674 (1979).
25. J. A. Chandler, X-ray microanalysis in the electron microscope, *in* "Practical Methods in Electron Microscopy" (A. M. Glauert, ed.). North-Holland Publ., Amsterdam, 1977.

---

[8] This list should by no means be considered to be comprehensive; rather, most of the references are used as sources for some figures or special discussions. The same is valid for the reference lists of the other chapters.

26a. M. von Heimendahl and H. Oppolzer, *Scripta Met.* **12**, 1087 (1978).

26b. M. von Heimendahl and G. Maussner, Journ Materials Sci. **14**, 1238 (1979).

27. D. R. Beaman and J. A. Isasi, Electron beam microanalysis, *Mater. Res. Std.* **11**, No. 11 (1971).

28. D. R. Beaman and L. F. Soloski, *Anal. Chem.* **54**, No. 9 (1972).

29. A. Hendriks, *EDAX EDITor* **5**, No. 3, 13 (1975).

30. J. C. Russ, *in Proc. 6th, Ann. Scanning Electron Microsc. Symp., Chicago, Illinois* p. 113 (1979).

31. C. Cliff and G. W. Lorimer, *J. Microsc.* **103**, 203 (1975).

2. Preparation

This chapter deals with the most important preparation procedures for transmission samples (necessary thicknesses of the order of magnitude 0.1 μm = 100 nm) from the field of materials research. Basically, two cases have to be distinguished.

a) The sample to be investigated is prepared directly to the necessary thinness (Sections 2.1 and 2.2).

b) A replica thin enough for transmission is made of the bulk sample surface to be investigated (Section 2.3).

2.1. Production of Thin Foils by Electrolytic Polishing

2.1.1. Prethinning to a Thickness of 0.15 mm

The technique of producing the necessary thin sample areas by electrolytic polishing requires as an initial product a metal sheet 0.10–0.20 mm thick, preferably 0.15 ± 0.02 mm. (The electrolytic thinning of thicker sheets takes too long, for thinner sheets the danger of bending is too great.)

For many experiments, e.g., examination of precipitation, recrystallization, cold-work, etc., the 0.15 mm sheet is best produced by *rolling* before the final heat treatment. If the fine structure in thicker materials, such as cast structures, is to be investigated, the necessary 0.15 mm metal sheet has first to be prepared from the material ("prethinning"). Several techniques are available. The simplest and most common is *grinding;* however, several precautions have to be taken. First, a disk of approximately 0.8–1.0 mm thickness and, for example, of 20 mm diameter can be produced by turning, sawing, cutting, or other mechanical techniques. This disk then has to be carefully ground to 0.15 mm from both sides to be absolutely *plane parallel*. This is difficult to achieve without the use of a *template* of tempered steel (Fig. 2.1) which on one side has a recess of 0.15 mm to hold the sample to be ground.

With a suitable, easily dissolvable adhesive, for example, Epoxy, the disk is glued into

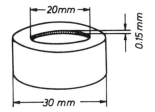

Fig. 2.1. Template for grinding, made of hardened steel.

the recess of the cylindrical holder. (Follow the instructions for the adhesive carefully). It is ground wet from one side to the desired thickness. For the last grinding step a 600 grit abrasive paper is used to produce a flat and smooth surface. The adhesive is dissolved with a suitable solvent, the sample is reversed and glued back into the holder with the side already ground facing down. The protruding part of the sample is then ground away until it is flush with the tempered steel rim of the holder. Again, it is important to use the 600 grit paper for the final step. To be able to remove the polished 0.15-mm thin, plane-parallel sample from the holder the whole assembly has to be soaked in the solvent for several hours. The disk can then be removed from the holder with a pair of tweezers without risk of bending it. For Al alloys, approximately 5–10 min are required to reduce the thickness by about 0.4 mm, steel with its greater hardness takes correspondingly longer.

The above-mentioned grinding procedure is suitable for all samples which are not too soft. Subsequent electrolytic polishing removes the surface layers which were possibly deformed during grinding. Only for extremely soft materials such as, for example, soft-annealed ultrapure copper or aluminum, this process is not applicable since the deformation due to grinding will go deeper than 0.07 mm. In these cases more careful thinning techniques have to be used; for example, spark erosion, "acid saws" (a nylon or metal thread which is dipped at one end into a suitable acid and, by moving back and forth, "saws" off the desired thin disk free of deformation), or an acid jet process. In the latter case, one has to distinguish between the cutting of thin disks with an electrolytic jet directed from a nozzle (details in Hirsch *et al.* [1] and Thomas *et al.* [2]) and the process of Theler [3] which consists of a combination of prethinning and final thinning. Common electrochemical or chemical polishing in most cases is unsuitable for the prethinning to 0.15 mm, since wavy and not plane-parallel surfaces are often formed.

Another very helpful piece of equipment has recently become available, namely a low-speed saw suitable for cutting metals and nonmetals alike. (No. 11–1180 Isomet low-speed saw by Buehler Ltd.) Such saws produce cuts low in deformation and work according to the principle of a circular saw. Approximately 0.2-mm-thick plane-parallel disks can be obtained which can be used directly for subsequent electrolytic polishing. The cut surfaces are no more severely deformed than those obtained by the grinding procedure described in Section 1.1. The cutting procedure may take up to several hours in larger samples but the saw stops automatically after completion of each cut.

2.1.2. Window Method

This is the simplest and most universally used electrolytic procedure which has the advantage of not requiring any special apparatus. In an electrolytic bath the sample is the *anode* placed, for example, in the center of a beaker (approximately 400 cm³), the *cathode* being a thin sheet of stainless steel lining the wall of the beaker (Fig. 2.2). A beaker of stainless steel is also suitable (easily cleaned, accident proof). Voltage may be supplied by any rectifier which delivers up to 60 V with a current load up to 5 A. Stabilization (as, for example, with potentiostatic methods) is not necessary when using the window method.

As is well known, electrolysis in such a cell is characterized by a current–voltage curve with a plateau in many, but not all, cases (Fig. 2.3). This plateau contains the "polishing region"; in the process of electro-chemical dissolution of the anode (sample), all microscopically protruding areas are dissolved preferentially, the end result being a shiny flat surface. Simultaneously the sample becomes thinner and thinner. If the voltage is too low the electrolyte produces an etching action (Fig. 2.3), if the voltage is too high pitting occurs. Both are unwanted. In electrolytes without a "plateau," the voltage suitable for polishing can be empirically determined.

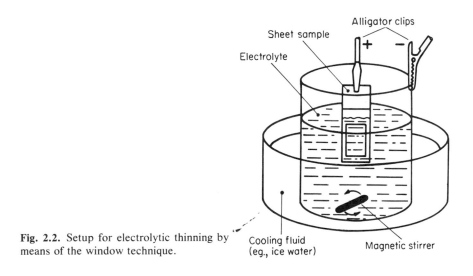

Fig. 2.2. Setup for electrolytic thinning by means of the window technique.

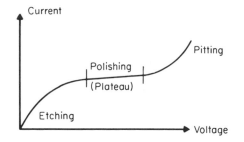

Fig. 2.3. Current–voltage curve for electrolytic polishing.

If, for example, a sheet metal strip with initial dimensions of 0.15 × 15 × 30 mm is used as a sample, as in Fig. 2.4a, it would start dissolving at the corners and the edges due to peak dissolution, without being "thinned." Therefore, one has to *insulate* the rim with an acid-resistant *lacquer* (Fig. 2.4b). Thus, a "window" is created and thinning occurs only in this area. At the upper edge a section which is not immersed in the acid is left unlacquered; here the sheet metal is connected to a support with an alligator clip or crossed forceps. The window-shaped area becomes thinner due to the polishing process until somewhere at the thinnest point a hole appears, Fig. 2.4c. The wedge-shaped edge of this hole often has thin areas of 100–300 nm thickness. This is enough for aluminum to be imaged with 100 kV electrons; not so, however, for heavier metals such as steel. To achieve the thickness of 50–150 nm necessary for steel, the edge of the first hole has to be coated again with about $\frac{1}{2}$ mm of lacquer (Fig. 2.4d) and polishing has to be continued until another hole appears. This process may have to be *repeated several times*. At the rim of the last hole sufficiently thin areas will be found.

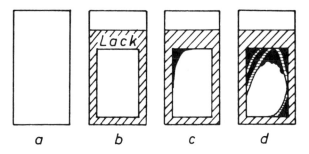

Fig. 2.4. Sheet sample as used for the window technique. Shaded = laquer, black = holes Lack = laquer, Natural size.

Sometimes the holes in a vertically hung foil form either only at the top edge or only at the bottom edge of the window. In these cases, it is advisable to turn the whole foil 180° after each relacquering so that top and bottom are exchanged. Electrolytes are of two basically different kinds; those which create a viscous surface layer which flows downward, thus the upper part of the sample is preferentially polished; and those which form bubbles on the surface. The bubbles rise and protect the upper part of the sample surface. In this case the bottom is more rapidly thinned.

When the sample has enough areas sufficiently thin, it is rinsed *thoroughly under running water,* rinsed in alcohol, and dried with a hot air drier. If the sample is not yet thin enough, the process has to be repeated or the Bollmann technique (Section 2.1.3) must be used.

Next, the sample is placed onto a plexiglass plate and a piece is cut off at the edge (as shown at the top of Fig. 2.5) with a scalpel *without deforming the sample* (best done under a good binocular magnifier). The knife should not be moved back and forth, but the cut should be made by using pressure only. The foil pictured in Fig. 2.5 shows at the top a frayed thin edge in contrast to the straight, cut edges. With the aid of a watchmaker's tweezer or vacuum tweezer, the severed piece is placed on a grid which in turn has been previously placed into the object cartridge of the EM. Examples of grids with different mesh sizes are shown in Fig. 2.5. To secure the sample foil, a ring is placed on top (Fig. 2.5 right) which, by screwing the cartridge halves together, is pressed against the sample and the grid. The ring locks the edge of the foil in position and leaves the center free for electron beam transmission. If the foil pieces

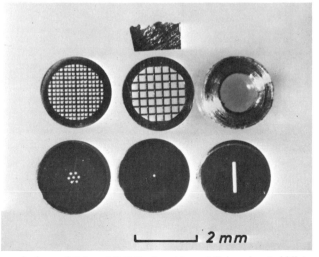

Fig. 2.5. Severed piece of thinned foil (top), grids and fixing ring (middle), and various specimen support disks (bottom) (8.3 ×).

are very small, they may be placed between two grids. Thus, the sample is ready for EM examination.

A selection of the most important electrolytes and polishing conditions for a few metals and alloys is given in Table 2.1. Very detailed tables containing 154 different recipes for many metals previously examined can be found, for example, in Hirsch *et al.* [1]. Some electroyles have to be heated or cooled and for almost all *bath stirring* is recommended, best by means of a magnetic stirrer. These are available on the market in combination with a heating plate. For all electrolytes with a working temperature below room temperature, the rule applies that the lower the temperature, the slower the polishing process and often the better the quality. Therefore, for example, a lower temperature is often used rather than the indicated temperature of ~4°C (ice water) for electrolytes Nos. 1 and 2 in Table 2.1. By the way, the *polishing voltage* is the primary and more important quantity affecting the polishing conditions, whereas the *current and current density* are variables depending on the sample size.

Furthermore, working with *perchloric acid* ($HClO_4$) is often thought to be dangerous. However, it should be noted that accidents (explosions), which have happened and which have been described in the literature ([24], p. 52), were *not* caused by the acid mixtures of Table 2.1. They occurred in electrolytes of perchloric acid, water, and *acetic anhydride*. Experience has shown that the electrolytes given in Table 2.1 are no more dangerous than acids generally, provided the following *precautions* are taken. During mixing, the $HClO_4$ has to be slowly and cautiously added, either drop by drop or as a thin flow. At the same time the mixture has to be kept *cool* (ice water bath). It is best to use a thermometer as a stirrer since this enables the user to make sure that the temperature of the mixture does *not exceed 10–15°C during the preparation*.

All new mixtures based on *acetic acid* tend to release gas during the first 10–30 hr. Therefore the bottles should not be tightly closed (leave stopper or cap loose). During electrolytic polishing an electrolyte based on acetic acid may be at room temperature, however the temperature should *not rise above 25°C* (otherwise, cooling is required). Open flames in the vicinity of $HClO_4$ are absolutely forbidden.

If these rules are followed the electrolytes given in Table 2.1 can be used unhesitatingly. About the use of $HClO_4$ generally consult the book by Tegart [24, Chap. 6].

Success of electrolytic polishing means obtaining uniformly thin, and at the same time clean, foils extending over large specimen areas. In some cases, this demands a great deal of patience, especially for alloys which have not yet been investigated. Deformed or multiphase alloys are more difficult to polish than undeformed or homogeneous metals or alloys. If a second phase exists in the form of larger inclusions (larger than 0.3–0.5 μm), the difficulty often arises that they are dissolved preferentially by the electrolyte and leave only holes (different electrochemical potentials). In such cases different electrolytes have to be used that polish both components more uniformly. The larger the size of the inclusions, the more difficult is the solution of this problem. On the other hand, particles of 1 μm diameter or larger are unsuitable for the EM, since these can be resolved in the light microscope. Extensive experience exists concerning the thinning of aluminum alloys. Single-phase alloys or those containing only very small particles are best polished by electrolytes Nos. 1 or 2 of Table 2.1. Both have the advantage of not leaving any troublesome oxide layers on the foil surface. However, for the above-mentioned reasons, they do not work well in some cases when larger precipitates are present. Electrolyte No. 3 (after Lenoir) avoids preferential dissolution and has the additional advantage of producing especially large and uniformly thin foils. On the other hand, it has the unfortunate disadvantage of leaving a porous oxide layer on the sample. These oxide layers must be removed after thinning by means of a suitable "washing solution" (Table 2.2).

To remove the oxide, the finished foil (Fig. 2.5 top) is held with tweezers and suspended in the hot solution for the prescribed length of time. Subsequent rinsing should *not* be done under running water since the tiny piece of foil would be deformed. Instead, the foil should be carefully dipped in a beaker with water. Finally, the foil is rinsed in a dish filled with pure alcohol and then dried on a sheet of filter paper. The effectiveness of the washing solution depends on the alloying metal, since on occasion the solution also dissolves the larger precipitates. Solution B is more effective in removing the oxide film; however, it has a greater tendency towards pitting and destruction of the sample.

Apart from the oxidation of aluminum, difficulties occasionally arise in the form of contaminated sample surfaces. At first one should always suspect the water and alcohol used in the final rinses. Generally, tap water is sufficient; distilled water is only needed in exceptional cases. The alcohol can normally be a commercially pure ethyl alcohol, also in denatured form (but *no* spirit alcohol). Should this fail to produce clean foils, the purest undenatured alcohol (pure ethyl alcohol) must be used.

Another source of failure in polishing may be an old (too often used) electrolyte which has been greatly enriched with ions. Finally, cases have been observed [4] in which specific ions have precipitated anodically in the form of nodules on the sample surface. In such cases, the specific combination of electrolyte-sample material is unusable. This is especially true of some electrolytes for precious metals.

All of the metal and alloy foils used as *examples* for the transmission electron micrographs

Table 2.1 Some widely used Electrolytes for Metals and Alloys

Material	Electrolyte	Voltage (V)	Current density (A/cm²)	Temperature (°C)	Remarks
1) Aluminum, Al alloys	65 cm³ methanol 33 cm³ HNO₃	8–10	0.5–1	≈4°C (ice water cooling)	Preferred for pure Al or single-phase Al alloys with GP zones or only small precipitates (inclusions). AlCu alloys with larger precipitates also polishable.
2) Aluminum, Al alloys	20 cm³ perchloric acid 80 cm³ ethanol	12–25 initially more, then less	0.2	<20 (tap water cooling)	While mixing, add perchloric acid carefully to ethanol (cooling!).[a] AlAg with larger precipitates also polishable.
3) Aluminum, Al alloys	393 cm³ H₃PO₄ (d = 1.7) 120 cm³ H₂O 80 cm³ H₂SO₄ (conc.) 93 g CrO₃	15	0.2–0.4	70	Electrolyte according to Lenoir, especially for Al alloys with larger precipitates (see text). Mixing: first dissolve CrO₃ in H₂O, then add H₃PO₄, finally H₂SO₄ slowly (cooling).
4a) Iron, steel, Cu alloys	675 cm³ CH₃COOH (glacial acetic acid) 125 g CrO₃ 35 cm³ H₂O	50–60	0.1–1	10–20	Preferred for Bollmann technique (see Section 2.1.3). Standard electrolyte for steels. Add acetic acid last while cooling.

Material	Electrolyte				Remarks
4b) Iron, steel, nickel, Zircaloy	10% perchloric acid 90% acetic acid	12–15		≤20	Standard electrolyte for all steels. Observe precautions in using perchloric acid.[a]
4c) Stainless steel	60% H_3PO_4 40% H_2SO_4	20–25	1–2	20	
5) Copper, CuZn	65 cm³ methanol 33 cm³ HNO_3	2–6	0.05	≤ –30	For high purity Cu cover electrolyte with liquid nitrogen. Wash in methanol (–30°C), then alcohol, take TEM micrographs immediately (to prevent oxidation). Also useful for thinning of wires [7].
6) Copper	50% H_3PO_4 50% H_2O	3		20	
7a) Nickel, Ni alloys	40% H_3PO_4 35% H_2SO_4 25% H_2O	6	2–3	20	Stirring not necessary.
7b) Nickel, NiAl alloys	5% perchloric acid 95% ethanol	18–20		0	Observe precautions in using perchloric acid.[a]
8) Silver, Ag alloys	6–9 wt% KCN in H_2O	6–10		20	Greatest caution necessary in using KCN (extremely poisonous!). Also, do not inhale the vapours. Study first-aid beforehand.
9) Titanium, Ti alloys	10% perchloric acid 90% methanol	6		–30	Window technique (also for wire preparation). Add perchloric acid carefully (cooling) and observe necessary precautions.[a]

[a] See special precautions concerning work with perchloric acid.

Table 2.2 Washing Solutions for Aluminum Foils, which were Thinned with Lenoir's Electrolyte No. 3

	Composition	Time and temperature of application	Suitable for the alloys named
Solution A	35 cm³ H_3PO_4 (d = 1.71) 20 g CrO_3 940 cm³ H_2O	10 sec, 70° C	AlAg, AlCu, AlSi, AlMgSi
Solution B	35% H_3PO_4 (d = 1.71) 10% HNO_3 (d = 1.4) 55% H_2SO_4 (d = 1.84)	30 sec, 60°C	AlCu, AlAu (in some cases may be diluted with H_2O prior to use; dilution 1:2 or 1:3).

in this book were thinned by the window technique (or the Bollman technique described below), except as specifically noted. (For example, the samples shown in Figs. 1.4, 1.10, and 1.13 were all thinned using the window method with electrolyte No. 1 of Table 2.1.)

2.1.3. Bollmann Technique

In contrast to the light metals, even repeated use of the window method does not always produce sufficiently thin foils in steels and other heavy metals (Fig. 2.4d). For these, the so-called Bollmann method is a distinct improvement. In contrast to the window method, however, a special apparatus is necessary as shown in Fig. 2.6. The basic idea consists in the use of a pair of precisely and easily adjustable point cathodes (of stainless steel) which concentrate the field in the electrolytic bath at any desired place in the vicinity of the sample.

The method consists of three steps:

a) In the beginning, the window method is repeatedly applied until the condition shown in Fig. 2.4d is reached. For this stage flat cathodes are used, or point cathodes with a large distance between point and sample (1–2 cm). This initial step of the Bollmann method is shown schematically in Fig. 2.7A(a). The edge formed last is covered with lacquer.

b) With the aid of suitable clips the pair of point cathodes[1] is now placed symmetrically at a distance of 1–1.5 mm next to the highly polished sample foil at a point which is approximately equidistant from all edges.

---

[1] Figure 2.6 shows the flat cathodes in position in the polishing cell while the point cathodes are on the table next to the apparatus. The point cathodes must be insulated with lacquer to the very tip ($\frac{1}{2}$ mm) to make full use of the field concentration at the tip.

Fig. 2.6. Apparatus for Bollmann technique.

After the current has been switched on, a small hole should form directly in front of the points (Fig. 2.7A(b)). If the hole does not form directly in front of the points, but again at the edge, then the points were not close enough to the sample.

c) As soon as the hole has appeared in the center, polishing is interrupted and the points are placed again at the *greater distance* (approximately 1 cm) or are exchanged with flat cathodes. Continuation of the electrolysis produces a new (second) hole at the edge of the previously polished remaining area in the window. (The necessary second hole at the edge may also be produced, as was the first, by the point cathodes). At the same time, the center hole will be enlarged. Both holes grow towards each other. The current has to be cut off *exactly* at that moment when either the two holes are barely separated by a very narrow bridge or after both holes have just merged so that two sharp "tongues" have formed at the point of merger (Fig. 2.7A(c)). The thinnest areas are those on the "bridge" or on the tips of the "tongues." Concerning the electrolytes to be used, the rinsing, etc., all explanations given in Section 2.1.2 are applicable.

The final polishing stages in opaque electrolytes, such as Nos. 3 and 4a in Table 2.1, can be checked only by withdrawing the sample quickly at short intervals. In order not to damage the sample with the closely placed point cathodes, it pays to mount the sample holder and the cathode holders together on a swing-up plastic plate. This plate can also be used to make the necessary precision adjustments of the point cathodes while they are outside the electrolyte. The supporting arms for the anode (sample is held by an alligator clip on a metal rod) and cathodes are shown in Fig. 2.6. Adjustments in all three dimensions are made by set screws. The cathodes are connected in parallel. The bath is agitated with a glass stirrer (driven by a small motor via a rubber band).

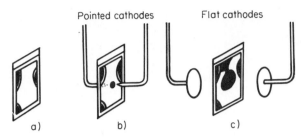

Fig. 2.7A. Steps during the Bollmann technique (see text). Black areas = holes, in stages a) and b) again covered with laquer (except middle hole).

2.1.4. Jet-Polishing Methods

Jet-polishing methods are relatively new developments which produce the best results by far. The yield often is close to 100% and the specimens contain especially large transparent areas (e.g., 10 μm edge widths). Polishing times are much shorter; however, specialized, though commercially available, apparatus are necessary.

The starting material is a prepared sheet, 0.10–0.25-mm thick. From this sheet disks are produced of a diameter which will fit into the specimen holder of the EM, i.e., 2.3 or 3.0 mm diameter. The disks are either punched or, better yet, cut without deformation by spark erosion.[2] Each disk is mounted in an insulating holder (e.g., Teflon) which exposes only the center portion of the disk to the electrolyte. Electrical contact with the disk is made with a platinum wire. To prevent electrolytic attack, the holder is designed so that the contact point between wire and disk is not exposed to the electrolyte. Jet sprays of electrolytic solution are directed against the disk through nozzles (Fig. 2.7B). Since these also serve as cathodes, they are made either of stainless steel, glass, or Teflon with an embedded platinum wire. In asymmetrical procedures, samples in a horizontal position are sprayed from the bottom, i.e., by only one jet [5]. In symmetrical procedures, a vertically positioned sample is sprayed with two jets, one on each side. In this method, the jets can either operate in air or submerged in the electrolyte [19]. The latter technique is particularly effective.

Again, it is of greatest importance to catch the *precise* moment when the hole first appears. Only at that very moment the edge of the hole

---

[2] In a circulating dielectric liquid (for cooling), high voltage sparks are generated across a narrow gap between the sample material and a template. The sparks erode the material in the desired shape. For details see the literature of the equipment manufacturers (for example, Servomet by Imanco, and Agietron by Vorkart).

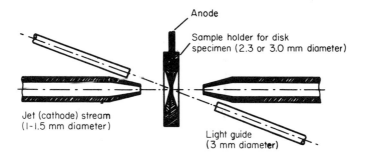

Fig. 2.7B. Jet-polishing technique (schematically).

contains the desired, very thin, transparent areas. With transparent electrolytes (see Table 2.3), a reliable method [19] is available. By means of a fiber glass light conductor ("light pipe"), a light beam is directed through the acid directly onto the center of the disk sample, Fig. 2.7B. Opposite,

Table 2.3 Electrolytes for Jet Polishing

Material	Electrolyte	Voltage (V)	Current (mA) for 2.3 mm disks
1a) Fe, steels	10% $HClO_4$ (perchloric acid) (60%) + 90% CH_3COOH (acetic acid) (96%)	12–13	20
1b) Ni, NiAl, Nimonic	same as 1a)	30	65
2) Zircaloy	same as 1a)	15	
3) Aluminum, Al alloys	methanol + HNO_3 (65%), 3:1 to 6:1	10	80
4a) Copper, Cu alloys	$H_3PO_4 + H_2O$, 1:2	6	80–90
4b) Copper	methanol + HNO_3, 2:1; wash in methanol and ethanol at room temp.	~6	80
4c)* Copper, α-brass	125 cm³ H_3PO_4 125 cm³ ethanol 250 cm³ H_2O 1 cm³ "Vogel's Sparbeize"	6–8	40–60
5) Titanium, Ti alloys	90% methanol 10% perchloric acid temperature: −30°C	~10	

* Keeps only one day.

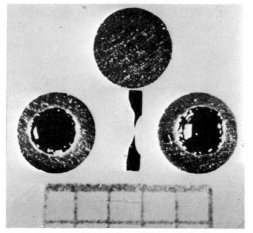

Fig. 2.7C. Disk specimen before and after jet thinning [30]. Plan view and cross section. Millimeter scale for comparison. Final hole (middle and right) less than 0.1 mm diameter.

a second light pipe is symmetrically placed, with a photo cell at the end, above the acid bath. At the moment of breakthrough when the hole has a diameter of *less than 0.1 mm,* the thin light beam reaches the photo cell which by means of a relay, shuts off the polishing voltage. This automatic cutoff practically guarantees success. Also, an acoustical signal informs the otherwise uninvolved operator of the finished preparation.

Figure 2.7C shows disk samples thinned in this manner, in plan view and in cross section. They were prepared in a twin-jet electropolisher, type Fischione. The dependence of the cross-section profile of the disk (center of Fig. 2.7C) on current density has been examined in DuBose and Stiegler [20]. If the current density is too high, the center of the disk remains thicker than the surrounding areas.

The jet polishing methods using disks have the following advantages over the window or Bollmann techniques:

1) Considerably *larger thin areas* are transparent.

2) The preparation is much *more rapid,* i.e., 3–5 min versus 15 min for aluminum and up to several hours for heavy metals.

3) *Smaller pieces of sheet* are needed for prethinning, consequently material is saved. Also, wires or foils of ≥2.3 mm in cross section can be examined.

4) Since the hole often appears close to the disk center (as in Fig. 2.7C), *aimed thinning* is possible (e.g., examination of solder joints).

5) *Lacquering* of the edges is *no longer necessary,* since the plastic material of the sample holder protects the disk edge during polishing.

6) *Grids are no longer necessary* since the disk is self-supporting (thin areas are not covered by grid bars).

7) The disk edge which remains thick (Fig. 2.7C, center) is a good heat conductor and protects against inadvertent mechanical damage of the thin edge of the hole.

8) Plane-parallelism or the absence of surface waviness are no longer critical during prethinning (compare Section 2.1). Therefore, starting materials may be thicker (for example, 0.5 mm) and can be prethinned either chemically or with the window method. This is a great labor saver.

9) Depletion and heating of the electrolyte proceeds approximately 100 times slower than with the window method due to the smaller surfaces being polished. Therefore, the electrolysis can almost always be done at room temperature.

10) The fact that the hole often appears precisely in the center of the disk is very useful if the sample is to be tilted in the EM (sample area will shift less during tilting).

11) The separate constituents of heterogeneous samples will be thinned more uniformly than with the window method.

For an example of aimed thinning ("spot preparation") by means of jet polishing, see v. Heimendahl et al. [20a]. To reveal the detailed structure of so-called transformation bands in a creep-resistant Al-Cu-base alloy it was necessary to hit specific sample areas seen in the light microscope with an accuracy of about 50 μm.

At present jet-polishing apparatus is available from the following companies:

a) E. A. Fischione, 7925 Thon Drive, Verona, Pa. 15147 (twin-jet electropolisher).

b) Buehler, Ltd. (AB-Electromet thinning apparatus, Cat. No. 70–1750).

c) Müller KG., D-5930 Hüttental-Weidenau, Postfach 248, West Germany (thinning apparatus according to Maurer and Schaeffer (Leoben)).

d) Struers, Inc., 20102 Progress Drive, Cleveland, OH 44136 (thinning apparatus Tenupol).

Table 2.3 is a list of electrolytes and polishing conditions which have proven helpful with the Fischione apparatus.

2.1.5. Preparation of Wires

Compared with the preparation of foils already dealt with, wire is no more difficult if longitudinal strips with minimum width of 1–3 mm and length of >10 mm can be produced from it. With a bit of ingenuity, these can still be thinned by the window or Bollmann method. The edges, however, have to be lacquered especially narrowly (under a binocular microscope) so that only a small area is lost. Wires with a minimum diameter of the size of the microscope grid (2.3 or 3.0 mm) may be prepared in the form of disks in any plane and jet polished according to Section 2.1.4.

If, however, the wire diameter is less than about 1 mm, individual wires can be thinned only with greatest difficulty with the procedures given. A breakthrough occurred with the idea of producing an artificial "sheet" by wrapping *many* wires onto a metal frame and then uniting them by electrolytic deposition of metal (electroplating). After careful plane-parallel polishing, such a "sheet" can be further thinned like any solid sheet. Either nickel or the same alloy as that of the wires is used for the electrolytic "embedding." Glenn and Duff [6] plated Cu, as well as steel wires of only 0.12 mm diameter, with nickel and were able to thin the "wire sheets" thus produced. For details see the original paper [6]. The most serious difficulty is that electrolytic attack often occurs during thinning, preferentially at the interface between wire and embedding metal, and there produces long holes, before areas in the wire interior have become thin enough.

A variation of this method was used successfully by Scheucher [7] for the preparation of copper wires (0.3 mm diameter). He used only *one single wire* on which he electroplated copper, in the form of a sheet. The thickness of the sheet was made equal to that of the wire by placing the wire, during the cathodic deposition, between two glass plates which were held in a parallel position by four short pieces of the same wire at the corners. Subsequent thinning was done with electrolyte No. 6 following a modified Bollmann procedure: a steel wire of 0.1 mm diameter served as the point cathode which was carefully aligned by hand under a binocular microscope so that the hole would form at the desired place. To enable the observer to see the hole formation, the recirculated electrolyte, in an approximately 1-mm-thick layer, is made to flow down an inclined plane on which the "wire sheet" is lying. The steel wire (cathode) is held first parallel and finally normal to the wire.

Figure 2.8 shows two pictures of samples which were prepared in this manner from 0.3 mm copper wires which were subjected, by a special procedure [7], to a short-term recrystallization anneal.

Rack and Cohen [21] have described a preparation procedure for wires without electroplating. Cross sections of Fe and Fe–Ti wires of only 0.13 mm diameter were implanted into a hole in a stainless steel sheet which was covered with lacquer. Thinning was done by jet polishing.

2.2. Production of Transparent Samples by Other Procedures

For metallic materials, the production of foils by electropolishing, as indicated in Section 2.1, is the prevalent technique by far. In very rare cases, other methods are used which shall now be discussed. These are

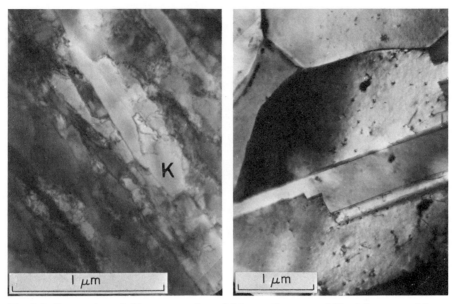

Fig. 2.8. Preparation of thin copper wires of 0.3 mm diameter., a) Hard drawn condition (90% cold drawn); small recrystallization nucleus (*K*) due to short-term recrystallization anneal, 50 msec at 350°C (40,000 ×). b) Fully recrystallized condition after 950 msec at 350°C (20,000 ×). (Micrographs: Scheucher [7] and D. Puppel.)

of greater importance for nonmetallic materials. All of these methods have one thing in common with the electropolishing procedures, in contrast to the replica technique of Section 2.3, namely that the material itself is *directly observed in transmission.*

2.2.1. Mechanical Procedures (Hammering, Cleaving, Cutting, Ion Thinning)

By *hammering* (forging), gold can be made so thin that it is easily transparent to 100-kV electrons (gold leaf). Figure 2.9 is an example of a piece of gold tinsel which was floating in the liqueur called "Danzig goldwater." Aside from washing the tinsel in distilled water and alcohol, no special techniques were used for the preparation of this sample. After the final alcohol rinse, the foil pieces are once more placed on a clean water surface in order to "smooth" them (compare Section 2.3.2). Figure 2.9 shows the numerous dislocation tangles and forest dislocations, as well as deformation twins, present after the severe cold deformation by hammering. The fact that deformation twins are present rather than stacking faults was proven by diffraction (Chapter 4).

Cleaving of metals is not possible; however, it is possible in some other crystalline materials, such as mica or graphite, which have a pre-

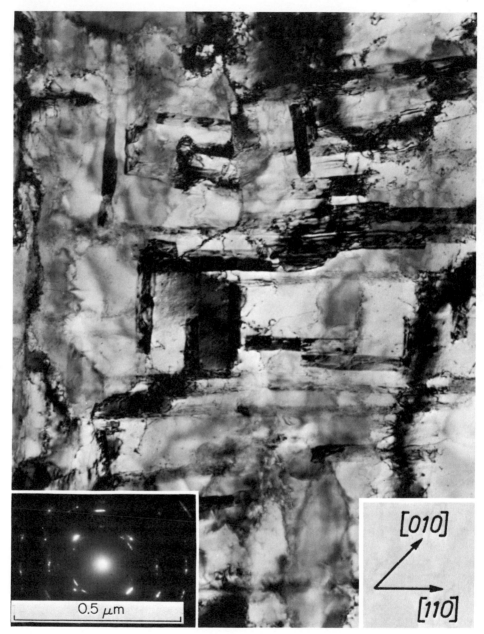

Fig. 2.9. Transmission electron micrograph of gold leaf. Due to the very high degree of cold work dislocation tangles and numerous deformation twins have developed. The twins appear as bars normal to each other in ⟨110⟩ directions, since the foil normal is [001]. For the diffraction pattern insert, see Section 4.7 (80,000 ×).

ferred *cleavage plane* due to their lattice structure. In such materials one can, on occasion, produce transparent lamellas by purely mechanical means. Thus, sufficiently thin *mica foils* can be prepared in the following manner. the mica sample is placed between two pieces of adhesive tape which are then pulled apart. This procedure is repeated several times until only an extremely thin mica flake remains on the tape. Small squares are scored into the tape (corresponding to the size of the microscope grids). By soaking the tape in distilled water, the mica foils are loosened, washed once more, and then fished out with small grids. Figure 2.10 shows a TEM photograph of a mica foil which was prepared in this way. The sample is already very thin, as evidenced by its good transparency. Nevertheless, it still consists of several superimposed crystal lamellas which give rise to a Moiré pattern (Chapter 4). At the edge of the sample, however, *single-crystal* lamellas were found without Moiré fringes. This investigation showed that the laminar structure of mica has dimensions accessible only to electron microscopy.

Moiré patterns consist of fine striae of close parallel lines covering the entire picture surface. These should not be mistaken for the much larger interference fringes S (Sections 1.6.2 and 4.5), although for contrast-theoretical reasons the Moiré patterns are especially distinct within the fringes (for example, at point M).

While hammering and cleaving are limited to relatively few exceptional cases, *cutting* with a *microtome* plays a greater role. In the microtome a rod-shaped sample is moved up and down while being advanced against a "knife," so that thin "slices" fall off (Figs. 2.11A and B). With an ordinary microtome, slices of the order of 1–5 μm can be produced, as are needed for light microscopy, while with an ultramicrotome, foil thicknesses of a few tens up to 200 nm can be produced, as necessary for electron microscopy.

The specimen block is tapered in front, as shown in Fig. 2.11A, since sufficiently thin slices can be obtained only from blocks of small cross section and since the diamond knives are very delicate. Most often the sample is moved up and down (with an adjustable thermal advance by means of incandescent-lamp heating) while the knife is stationary. This is advantageous because the knife is situated at the end of a small trough filled with liquid, e.g., water, containing 10% acetone or ethanol. This liquid serves to catch the freshly cut, extremely thin and very small sections which will float freely on its surface. The liquid has to extend, there-fore, directly up to the cutting edge of the diamond knife and must have the correct degree of wettability. If both these conditions are not met, there is danger the thin sections may roll up instead of remaining flat on the surface.

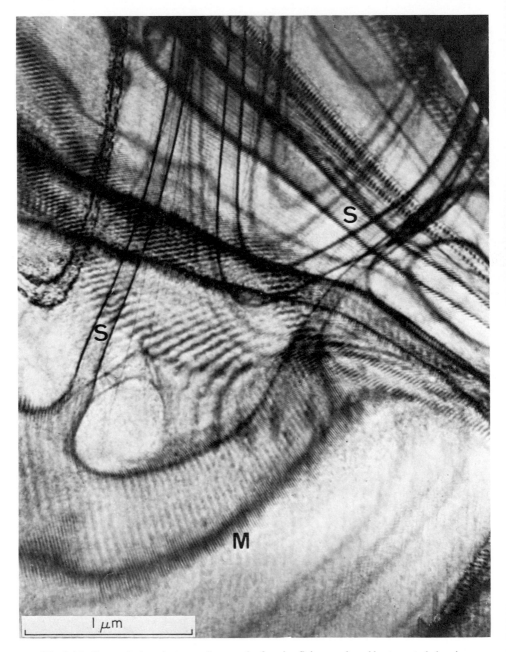

Fig. 2.10. Transmission electron micrograph of a mica flake, produced by repeated cleaving. Moiré pattern *M* and double interference fringes *S* (for interpretation see Section 4.5.4) (40,000 ×).

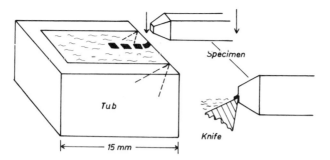

Fig. 2.11A. Principle of ultramicrotome cutting.

The thin slices follow each other cut after cut, so that their original location in the bulk sample remains unambiguously known. Thus, the effect of sample location or depth on specimen structure can also be investigated. The sections are fished out with fine-meshed support grids (400 mesh grids). The edge lengths of the thin slices are comparable to the size of the grid openings.

More details as to the construction and principle of operation of ultramicrotomes can be found in Reimer [8], as well as in the literature of the manufacturers.

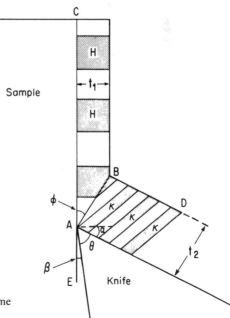

Fig. 2.11B. Processes during microtome cutting. See text and Phillips [9].

Although soft or medium-hard metals can be cut without great difficulty with an ultramicrotome, its use is not advisable because of the very severe deformations which occur. These will be superimposed on the originally existing microstructures to be investigated. In spite of this unwanted deformation of metallic samples, the ultramicrotome is used in certain exceptional cases when it is otherwise impossible to obtain thin, transparent samples. Several interesting cases of that kind are described in a paper by Phillips [9]. Figure 2.11B, taken from that paper, elucidates the process of the microtome "cut." The sample is moved along the knife (diamond for metals) in a downward direction, yielding a thin chip or shaving similar to those from turning or milling, so that the term "cutting" is not quite appropriate. The severe deformation during the shaving process is evident in Fig. 2.11B: The thickness t_2 of the shaving slipping along the knife is greater than the thickness t_1 removed from the sample. This means that the foil produced is *severely compressed*. Sample areas originally square (H) are *deformed* into parallelograms (K). Since this compression does not occur in the direction normal to the plane of the drawing (i.e., parallel to the cutting edge), the shaving is also *distorted*.

The following are examples for ultramicrotome applications:

a) Examination of small holes in porous samples—during electrolytic thinning the holes would be enlarged and therefore would not be representative of the original material. Figure 2.12 gives an example of pores in a Cu–3% Co foil [9]. The above-mentioned *distortion* in the direction of the arrow can be recognized.

b) Examination of sections normal to the plane of thin foils (for example, 0.1 mm thickness) (Fig. 2.12 depicts such a case).

c) Examination of corroded or oxydized surface layers normal to the layer plane. As an example, Fig. 2.13 shows a thin section through an anodic oxidation layer on an Al–1% Mg alloy sheet. The layer still adheres to the metal surface, although it is severely cracked due to its poor plasticity. In addition, hard inclusions are visible in the metal substrate. Unknown surface layers can be identified by SAD.

Figures 2.12 and 2.13 also show the above-mentioned compressive deformation.

Recently it has also become possible to section *glasses* using an ultramicrotome with a diamond knife. According to Krauth [10], glass may be "cut" with an ultramicrotome in the following manner: a needle-shaped sample, with a cross section of about 0.2 × 0.2 mm, is coated with a bonding agent (e.g., chrome methacrylate) and then embedded in a curable plastic (e.g., Araldite) which can be mounted in the microtome sample mount. At a cutting speed of 2.5 mm/sec, the sample is moved

Fig. 2.12. Application of the ultramicrotome: thin cut of a porous Cu–3% Co sheet (cut normal to the plane of the sheet) (2000×). (Courtesy Phillips [9].)

downward against the diamond knife and forward by thermal expansion, causing small shell-shaped lamellas of approximately 10 × 10 μm surface area and 50–100 nm thickness to spall off the glass. These lamellas spread out on the isopropanol-covered water surface of a trough which extends to the back side of the knife. If a specimen grid is brought close to the water surface, the glass lamellas jump on it by themselves.

Figure 2.14 shows a Cu–ruby glass that was prepared in such a manner. Note the very high magnification of the picture (250,000×) [10].

Plastics too can be cut thin enough with an ultramicrotome (diamond knife) to be transparent to electrons. As an example, Fig. 2.15 shows hard PVC with embedded crystallites of basic lead sulfate for stabilization.

A further possibility for electron microscopic examination of plastic consists in producing small transparent crystals by crystallizing them out of suitable solutions. In this manner, polyethylene crystals (PE) were prepared on a carbon support film by evaporation of a xylene solution [27]. Contrast effects were examined in more detail on similar PE crystals by Grubb *et al.* [28]. One complication arising during electron irradiation of plastics is the damage done by the beam electrons. In contrast to metals,

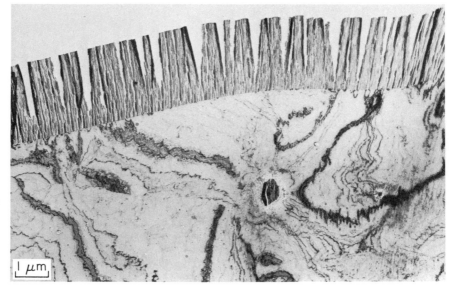

Fig. 2.13. Application of the ultramicrotome: thin cut across an Al–1% Mg sheet with an anodized surface (8500 ×). (Courtesy Phillips [9].)

beam damage occurs in plastics at voltages as low as 100 kV. These effects have been examined by Grubb and Groves [29].

In the broader sense, *ion thinning, ion sputtering, or "ion milling"* can also be included in "mechanical procedures."

Ion-thinning works as follows: The disk-shaped, mechanically pre-thinned sample is bombarded at a near-glancing angle with suitable ions (e.g., argon ions). These ions randomly knock out surface atoms and uniformly reduce the thickness of the sample. During the bombardment of (usually) both sides simultaneously, the sample is rotated around its surface normal to produce uniformly dished specimens. The angle of incidence generally is 75–85° to the disk normal, the accelerating voltage is 5–10 kV. The thinning process is monitored with an optical microscope. The appearance of a hole indicates completion. However, continuation of bombardment does not round the hole edges as it does with electrolytic polishing. Consequently, thinning can be continued if desired, to obtain different transparent areas from the same disk.

Ion thinning is especially useful for *nonmetallic materials* (such as glass and ceramics), if all other procedures fail. In that case, one must accept the disadvantages of ion thinning, namely, that thinning is very slow (e.g., 20–40 hr for removal of a 150-μm layer) and the equipment is relatively complex and expensive. The interested reader should consult the original papers, such as Bach [22], Barber [23], and Gillespie *et al.* [26], which describe apparatus and procedures in greater detail.

An interesting application of ion thinning for *metals* is described by Wolff [31]: Neutron irradiated samples of stainless steel contained numerous large voids (>200 nm) of which the exact size and shape was to be determined in the EM. Electrolytic thinning would enlarge the voids and round their edges, mechanical polishing would distort and possibly fill the voids with debris. Therefore, the sample was polished (but not to transparence) by ion thinning with 7-kV argon ions (ion current = 100 μA). Within 4 hr approximately 10 μm were removed. The polished surface was replicated with cellulose acetate tape from which secondary, Cr-shadowed carbon replicas were made. In the EM these showed the voids in their original shape.

2.2.2. Chemical Methods

Attempts to chemically thin metals with suitable acids (chemical polishing solutions [25]) usually fail due to nonuniform removal rates. For nonmetallic crystalline materials, however, where electrolytic polishing is not possible, chemical thinning is often the only way to produce a transparent sample (e.g., for MgO, spinel, NaCl, Ge, etc.).

The preparation of MgO is used as an example: Platelets of about 6 × 6 × 0.3 mm (as thin as possible), obtained by mechanically cleaving

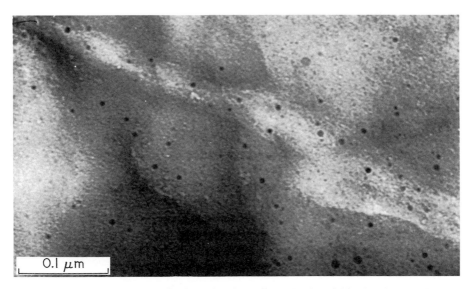

Fig. 2.14. Ultramicrotome cut of a Cu–ruby glass. The red color of this glass is caused by the very small spherical Cu precipitates (250,000×) (Courtesy Krauth [10].)

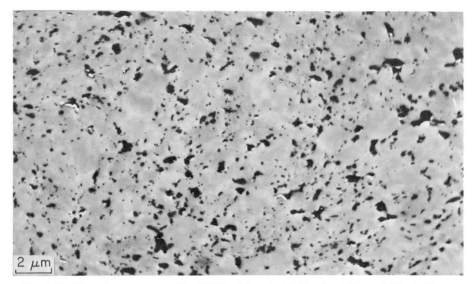

Fig. 2.15. Ultramicrotome cut of hard PVC with embedded lead stabilizers (4400×). (Courtesy Metall-Laboratorium der Metallgesellschaft AG., Frankfurt.)

a larger crystal, are used as starting material. A chemical polish for MgO is obtained with hot undiluted orthophosphoric acid (160–170°C). The temperature has to be close to the boiling point, since at lower temperatures (about 100°C) not polishing, but etching occurs. If the sample is just dipped into the solution, it would dissolve preferentially at the corners and edges. The lacquering of the edges described in Section 2.1 is replaced here by concentrating the acid action at the sample center with a *jet stream*. The hot acid jet is directed against the horizontally held MgO plate from a glass nozzle (diameter 0.3–0.5 mm) mounted 1–2 cm below the sample. The sample is reversed often to obtain a similar removal from both sides. *Immediately* after formation of the first hole, recognizable by the spray of acid penetrating the sample, the latter has to be removed and rinsed under *running hot* water, washed in alcohol as usual, and dried. Under a binocular microscope, the desired smaller sample pieces, which contain the thin areas at the hole edge, are isolated with a scalpel. As an example of an application, Fig. 2.16 shows a sample from the MgO–Cr_2O_3 system (as-cast, containing 3% Cr_2O_3). The rosette-shaped particles are the precipitated spinel phase $MgCr_2O_4$ in the MgO matrix material. The same thinning procedure can be used for spinel crystals (e.g., Al_2O_3/MgO); however, higher temperatures are necessary [11].

Transparent NaCl samples can be prepared in a similar manner. The cleaved platelets are polished and thus thinned at room temperature with a mixture of ethanol and methanol, to which some water can be added to accelerate the process (pure water acts too fast). NaCl, however, tends to be irreversibly damaged during the irradiation by the electron beam. Micrographs must therefore be taken immediately or under especially weak beam currents with an image intensifier.

For the chemical thinning of Germanium, see the original publication by Alexander [12].

In conclusion it should be mentioned that by *electrochemical deposition* or by *evaporation* of suitable metals, good transparent samples can be produced. Usually, the metal ions or atoms are deposited on a support film or substrate which, in turn, is easily dissolved as, for example, NaCl. The thin film floats on the surface and is then washed and dried. However, the specimens thus produced do *not* represent the bulk *"materials,"* but exhibit the very *specific features* of thin films. Therefore, this is not the place to go into details of this vast and interesting branch of electron microscopic investigations. The interested reader is referred to a comprehensive presentation, e.g., that of Mayer [13].

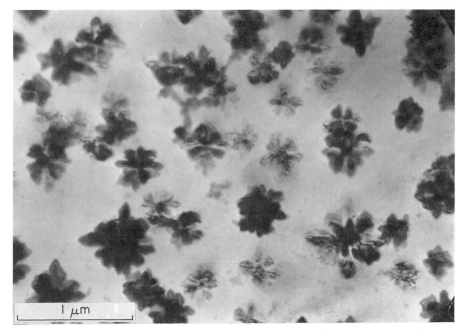

1 μm

Fig. 2.16. Example for chemical thinning. Material: MgO–3%Cr$_2$O$_3$. The rosette-shaped particles are the precipitated spinel phase (30,000 ×). (Micrograph: R. Willig and H. Steinbrecher.)

2.3. Replica Techniques for Surfaces

All preparation techniques discussed until now had in common that the material to be investigated could be *directly* prepared for electron transmission. In those cases in which either thinning to a size of 0.1 μm is *impossible,* or one is interested in the *surface of the sample* (e.g., studies of *fracture surfaces*), the replica technique can be used. The basic idea consists in creating a negative (replica) of the surface relief of the bulk sample, e.g., an etched metal surface. The negative has to be separable from the sample and has to be thin enough for EM transmission. A replica with such properties, and created in a *single-stage procedure,* is called a *film replica.*

Often, however, difficulties arise in stripping the film, especially with rough or strongly fissured surfaces: in such cases, a *two-stage procedure* is necessary. First, a thick replica (the so-called mold) of the sample surface is made with a plastic (either tape or bulk) which can be peeled off relatively easily without damaging it. Then, as in the single-stage process, a thin, transparent film replica is produced from the negative. The final replica thus is a positive of the original surface relief and is called a *two-stage mold replica.*

2.3.1. Single-Stage Techniques (Plastic, Carbon, Oxide Replicas)

Only the most important of the numerous tested methods will be described here. *Plastic replication* (Fig. 2.17a) consists of allowing a dilute plastic solution to dry on the surface, so that over "valleys" of the sample relief a thicker layer, and over "mountains" a thinner layer of plastic, is formed. By *carbon film evaporation* (Fig. 2.17b), a replica is produced which has approximately uniform overall film thickness in the vapor deposition direction. *Oxide replicas* (Fig. 2.17c) are created by an anodic oxidation process forming a film which has a uniform film thickness in the direction normal to the surface. Since the films are amorphous, the electron microscopic contrast is pure absorption contrast (also called structure factor contrast, compare Section 1.6.1), dependent only on the film thickness in the electron transmission direction. Darker regions represent thicker film areas. They all have in common, that *steps, "mountains,"* and *"valleys"* are represented by characteristic and discontinuous *thickness changes* of the replica film. The image contrast, however, in the "images" of the surface profile is different in all three cases.

Under the conditions pictured in Fig. 2.17, a groove-shaped depression is "imaged", in the case of the plastic replica as a darker strip (lower image brightness I_d), in the case of the evaporated (carbon) film as a pair of narrow lines, one light and one dark, and in the

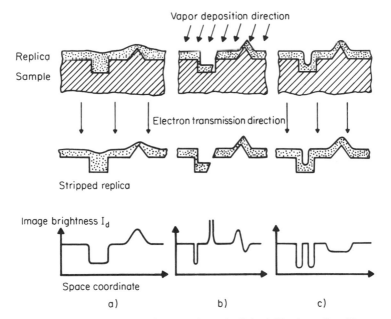

Fig. 2.17. Single-stage replica techniques (schematically). a) Plastic replica, b) evaporated film replica, and c) oxide replica.

case of the oxide replica as a pair of dark lines (see the indicated image brightness I_d in the lower part of Fig. 2.17). As can be seen, the replica techniques are very indirect procedures, and special care is necessary in interpreting the contrast.

Plastic replicas

The following are examples of commonly used plastic solutions of proven efficacy:

a) a solution of 0.1–0.5% collodion (pyroxylin, nitrocellulose) in amyl acetate,

b) a solution of 0.1–0.3% Formvar or Mowital in chloroform.

Formvar consists of polyvinyl formaldehyde and is a white powder. Mowital is a German trade name for the same substance. The highly diluted plastic solution is dropped from a pipette onto the sample, for example, an etched metal surface. The excess liquid is drained off the surface by tilting the sample. After evaporation of the solvent, one attempts to remove the plastic film from the surface with as little distortion or damage as possible. This is the more difficult the rougher the surface is. In rare cases, the plastic film will lift by itself after careful, inclined immersion into a water bath (it will "float off"), but occasionally it may do so after alternating baths in cold and warm water (thermal stress). One can also try pouring a *thick* layer (0.3–0.5 mm) of collodion on top of the Formvar film, as a so-called "reinforcement." After drying, this can usually be easily lifted mechanically together with the thin film (loosen it at the edge with a razor blade). The reinforcement layer is dissolved again in amyl acetate. Another technique is to pull the plastic film off with adhesive tape and then detach the latter in water. In all of these methods the sample surface to be examined remains undamaged, so that new replicas can be made in the event of failures. If a suitable replica cannot be detached after repeated trials, the remaining alternative is to use acids or solutions suitable to dissolve the sample or the sample surface without damaging the plastic film.

In all cases, it is recommended to score the plastic film beforehand with a pointed needle into small squares of the size of the specimen support grids or apertures (Fig. 2.5). These squares will float freely on top of the water after the last step of the stripping procedure. They can be transferred with a round glass rod or a small pipette into a second, clean water bath for washing. Finally, they can be fished out with a grid or a specimen aperture, brought up from below, and dried in air by carefully placing the grid (edge first) onto filter paper.

Figure 2.18 shows an example of a single-stage plastic replica. The Mowital film (B 30 H) was removed with Triafol as a reinforcement (refer to Triafol technique in Section 2.3.2) and then obliquely shadowed (see below) with C–Pt to increase the contrast. The Triafol was finally dissolved in acetone.

Evaporated carbon films

The most common substance for evaporated film replicas is spectroscopically pure carbon. Carbon layers have an *atomic* structure in contrast to plastics which are *molecular* (molecule sizes up to about 10 nm). Consequently, carbon layers have a smaller internal structure and therefore the ability to reproduce finer details, i.e., they have better resolution. Consequently, higher useful magnifications are possible for carbon films than for plastic films. In addition, with carbon films, there is less danger of distortions and other artifacts during, as well as after, the replication process.

The evaporation is best done in a commercial *evaporator*. Their function and the mechanics of evaporation are described in the operating instructions and manuals of the manufacturers. The necessary carbon rods (one pointed, one flattened) are made of spectroscopically pure carbon, as commonly used for arc lamps.

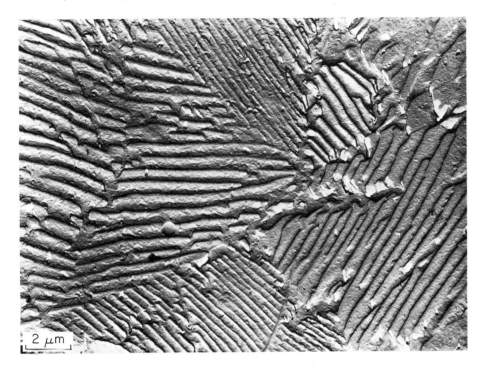

2 µm

Fig. 2.18. Single-stage Mowital replica of a pearlite surface. Contrast increased by oblique shadowing with C–Pt (6000 ×).

Figure 2.17b shows that the oblique evaporation necessary for contrast production creates "shadows" behind some edges, where the film could tear. To avoid this, evaporation is often done in two steps, each under a different angle (first oblique and then normal to the surface). Normal evaporation often yields insufficient contrast. To obtain a stronger contrast effect than by oblique evaporation of carbon alone (Fig. 2.17b), *heavy metals* are often used, e.g., platinum. A good method is to insert a platinum wire, (approximately 1 mm diameter) or a pressed pellet of mixed C and Pt powder into the predrilled center of a carbon rod. In this manner *platinum and carbon are evaporated simultaneously.*

For separating the carbon film from the specimen the same procedures can be used as were described for plastic films. From very smooth surfaces (e.g., glass or polished metals), the carbon film will float off without difficulty if the specimen is immersed obliquely into the water. From etched steel surfaces, the carbon film will loosen after repeated alternating immersion into water and 20% nitric acid in alcohol. It is recommended to let the sample with the film dry between immersions.

Another, important method for separating plastic or carbon films, especially for metals, consists of electrolytically dissolving the surface layer under the film. Any electrolyte suitable for electrolytic polishing of the metal (see Section 2.1) may bring about successful film separation, provided that the polishing takes place without gas evolution and without formation of insoluble particles. Finally, the film can be lifted by cathodic evolution of hydrogen, where the sample is the *cathode* [8]. Hydrogen bubbles creep under the scored film and loosen it.

Here, two nonmetallic samples are selected as examples of single-stage carbon replicas. Figure 2.19 shows etch pits at the exit points of edge dislocations on the surface of a LiF crystal deformed about 3.5% by compression [18]. The evaporation was done first obliquely with Pt–C and then in normal direction with carbon alone. The film was loosened with HBF_4 (solvent for LiF) and then washed in water and alcohol.

Figure 2.20 shows a single-stage carbon replica of paper (in this case

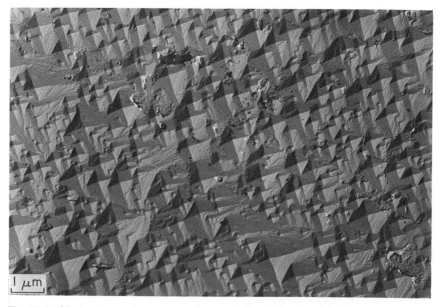

Fig. 2.19. Single-stage carbon replica of a (100) LiF surface containing numerous etch pits. The crystal was chemically polished with HBF_4 and then etched with a solution of 120 cm^3 HF, 120 cm^3 acetic acid, and 0.036 g Fe-III-chloride. Thus, an etch pit is formed at the exit point of each edge dislocation. (Mean dislocation density $N \approx 3.8 \times 10^8$ cm^{-2}, 8000×). (Micrograph: Reppich [18] and D. Puppel.)

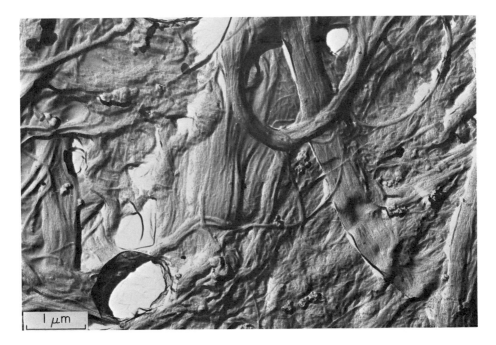

Fig. 2.20. Single-stage Pt–C replica of a "smooth" paper surface (16,000×).

glossy copying paper). One sees the typical structure of interlocking cellulose fibers of very different thicknesses. The evaporation again was
done obliquely with Pt–C and then normally with carbon. The film was
lifted by dissolving the paper in concentrated sulfuric acid.

Oxide replica method

This method, suitable especially for aluminum and its alloys, was frequently used before the direct electrolytic thinning of metals was known.
Aluminum surfaces can be easily oxidized by using, for example, a 3%
potassium citrate solution. The sample is the anode, the necessary voltage
is up to 20 V; the oxide layer formed is basically proportional to the
applied voltage and not to the duration of the process, since the growing
oxide layer blocks the passage of current. A sure way to remove the oxide
film is with saturated mercuric chloride solution ($HgCl_2$), which causes
a Hg film to seep between the oxide layer and the sample. Before the
customary rinsing in pure water, cleaning in dilute hydrochloric acid is
recommended.

When aluminum is etched chemically or electrolytically (e.g., with electrolyte No. 2 of Table 2.1, but at a lower voltage than given for polishing), removal of material takes place in such a way that cube faces, corresponding to the cubic crystal structure, are exposed on the surface. The resulting surface features are too small for light-microscopic observation. An oxide replica of this surface, however, shows in the EM many small rounded cube corners or faces, see Fig. 2.21. From the projected geometry, the crystallographic orientation of the crystallite (grain) or single-crystal surface under observation can easily be deduced. In the case of Fig. 2.21, the sample was a single crystal with a (111) surface. It had been chemically etched in a solution of 2 parts 40% HF, 3 parts HCl, and 4 parts H_2O (for 2 min). To enhance the contrast, the oxide replica had been shadowed obliquely at 45° with carbon.

Oblique shadowing to enhance contrast

In principle, shadowing can be used for all types of replica films, whenever the contrast is too low after the procedures described so far. Shadowing consists in evaporating onto the replica film a contrast material, usually a heavy metal, at an oblique angle of 30–60°. Usually Pt, or Pt–C simultaneously, is evaporated (see above). The "light and shadow effect," which is the basis of contrast formation, is illustrated in Fig. 2.17b. Plastic films can be shadowed only after being peeled off and turned over, because only the film side facing the sample has the true surface profile (see Fig. 2.22). The pieces of plastic film are prepared (fished out) on grids and then shadowed in the evaporator. With oxide replicas, on the other hand, oblique shadowing can be done before separation of the replica film from the sample.

2.3.2. Two-Stage Techniques (Technovit, Triafol)

As already mentioned, the fragile thin films frequently cannot be detached without being damaged, as happens, e.g., with etched metals. On the other hand, a solid plastic replica as a "mold" of the surface to be investigated can easily be lifted off. With the *Technovit method* the two-component plastic Technovit[3] is used. It sets within 15 min and can be lifted off fairly easily, although sometimes only after alternating hot and cold baths. The Technovit surface which contains the negative of the sample surface is then covered with an evaporated carbon film as de-

---

[3] Technovit #5071. Supplier: Kulzer & Co GmbH, 638 Bad Homburg vdH, PO Box 261, West Germany.

Fig. 2.21. Oxide replica of an Al crystal in (111) orientation (25,000×). (Micrograph: G. Vierling.)

scribed above. (Usually, first Pt–C is evaporated obliquely, then C normally, see Fig. 2.23A.) The carbon film, scored into suitable squares, is easily removed from the Technovit by immersion in an acetone bath (acetone dissolves Technovit). To clean the carbon film entirely of Technovit residue, the film must be transferred into one or two fresh acetone baths. At this point, the film pieces often contain numerous folds and distortions. These can easily be eliminated by transferring the squares onto a clean water surface. The high surface tension of water straightens out the films in one sudden movement. The final water bath has to be used in all cases where the last solution has a higher wettability than water as, e.g., alcohol. (If the straightening action of the water is too vigorous, breaking the films, the surface tension can be adjusted with a few drops of alcohol.) Finally, the films are lifted off the water surface with a grid brought up from below and dried by carefully placing the grid (edge first) onto filter paper.

Figure 2.23B shows an example of a Technovit replica made as de-

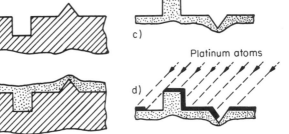

Fig. 2.22. The effect of oblique shadowing: a) sample surface, b) surface plus plastic film, c) plastic film peeled off and turned over, d) as c), but after oblique shadowing with Pt (contrast enhancement). (From H. Poppa.)

scribed. The *fracture surface* of polycrystalline Al_2O_3 is shown. The tablet-shaped sample was fractured in a rotationally symmetric bending test, then a drop of Technovit was placed on the fracture surface (approximately 1 mm wide). The fracture appearance differs locally. Some grains exhibit grain boundary surfaces as a consequence of *intercrystalline brittle fracture* as e.g., at the left edge of the figure. Others (in the center) show step fracture with numerous crystallographic fracture planes, some of them very small. This stepped fracture probably is a *trans-crystalline brittle fracture*. The micrograph was chosen to illustrate the ability of this replication technique to reproduce very fine detail.

In addition to molds of self-curing plastics, molds can also be made with temporarily softened, thicker plastic sheets or tape, such as those widely used in the electrical industry as insulation tape. Thus, for example, with the *Triafol method,* a 0.04-mm-thick (violet) foil is placed with the dull side on the sample surface which has been flooded with acetone. The acetone dissolves and partially liquefies the plastic surface. Thus, the plastic penetrates into all recesses of the sample surface. After 2–3 min, the foil is dry and can be stripped easily from the sample; it may come off by itself.

SiO is the preferred substance for evaporation onto the Triafol mold (use several 0.5–1 mm pieces in a tantalum boat). Evaporation is done first obliquely and then normal to the surface. However, C is also suitable for evaporation. If the Triafol is dissolved at this point, the fragile SiO or C film would be torn because of the swelling of the Triafol. Therefore, the evaporated film must be "reinforced" by an additional layer of paraffin, approximately 0.5-mm thick. This is done by pouring liquid paraffin onto the *free surface* of the C or SiO film. Now the Triafol film is dissolved in a bath of methyl acetate saturated with paraffin. The last Triafol residue

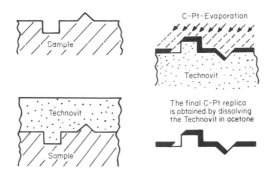

Fig. 2.23A. The 4 stages of the Technovit replication technique.

Fig. 2.23B. Technovit replica of fracture surface of polycrystalline Al_2O_3 ("step fracture", for details see text, 5200 ×). (Micrograph: H.-D. Schmidt and D. Puppel.)

is removed in a second such methyl acetate bath. The paraffin layer with the evaporated film is cut with a razor blade into squares with edge lengths of the diameter of the specimen grids (2.3 or 3.0 mm). Each square is placed onto a specimen grid, with the film side facing down, and the paraffin is dissolved by careful immersion of the loaded grid into a toluene bath. If a drop of toluene is placed on top before immersion, the film is prevented from slipping off the grid. A second toluene bath is necessary to remove the paraffin residue. Finally, the specimen grid with the finished replica is lifted from the toluene bath and dried on filter paper.

The Triafol method, because of the additional operations all of which include the danger of damaging the evaporated film or of producing artifacts, is more complicated and time consuming than the Technovit method. Occasionally, however, the use of Triafol has certain advantages in regard to sampling. This is true, for instance, for the often very porous surfaces of samples of glass or ceramics. The next three figures are taken from this field as examples of the capabilities of the Triafol method (according to Schüller [14–16]). The replicas were made by first evaporating chromium obliquely at 60–70° and then carbon normally.

The first two figures are from the field of *electroceramics*. Figure 2.24

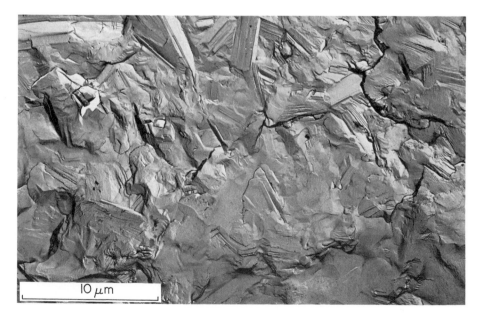

Fig. 2.24. Triafol replica of an unetched fracture surface of steatite (3600×). (Courtesy Schüller [14].)

shows the Triafol replica of an unetched fracture surface of *steatite*. One of the most important materials for low voltage insulation, it is also used in high voltage insulators. Steatite is produced from talc to which small amounts of clay are added to improve its processing properties. Another addition is feldspar or barium carbonate as a fluxing agent. In the fired condition, steatite consists of "Protoenstatite" crystals embedded in a glass phase [14]. In unetched fracture surfaces, the glass phase, which envelops the crystals and partially covers them, generally dominates the structural constituents in the micrograph. The striations in the crystals (see Fig. 2.24) are caused by the characteristic cleavage of the pyroxene on (110) planes. Thus, two identifying characteristics for the crystals are available: morphology and crystallographic features. ("Protoenstatite" belongs to the pyroxene class having the composition $MeSiO_3$ with a chain structure.)

The second example is the natural surface of polycrystalline *barium titanate*, Fig. 2.25. This has a threefold importance in materials technology: as the insulator material in condensers, as a piezo ceramic material, and as a ceramic semiconductor. For condensers, barium titanate is of interest because of its relatively high dielectric coefficient, so that large

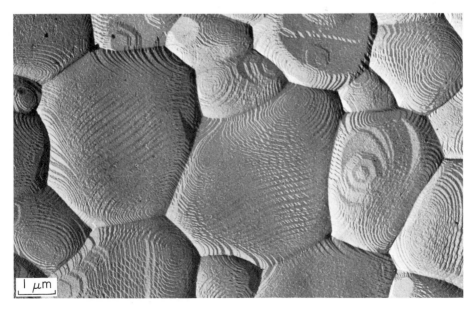

Fig. 2.25. Triafol replica of polycrystalline barium titanate (natural surface, 10,000×). (Courtesy Schüller [14].)

capacitances can be produced in very small spaces. Apart from the chemical composition of the material, the dielectric coefficient depends mainly on the microstructure of the dielectric. Normal barium titanate, produced as a ceramic, has a dielectric coefficient of 1200 at room temperature. If the compound is produced in such a way that it consists of only small crystals (1–5 μm), the dielectric coefficient is increased to 3000 at room temperature. The simplest way to obtain such small grains is with an excess of 4–8% TiO_2. With the optimal grain size of approximately 1 μm, the resolution limit of the light microscope is reached. A Triafol replica in the EM (Fig. 2.25), on the other hand, shows the grain boundaries very well (better in naturally grown surfaces than in fracture surfaces). Figure 2.25 also shows lamellar growth steps, which tend to form at free surfaces during growth of the crystals. Such growth lamellas, however, are typical not only for barium titanate, but also for other ceramic oxide materials (e.g., Al_2O_3).

The third example of a Triafol replica is taken from Schüller's [15, 16] extensive EM investigations of *porcelain*. Figure 2.26 shows the fracture surface structure of a common quartz porcelain, of which the four most important structural components are given in the figure caption. In contrast to the samples of the last two micrographs this fracture surface was etched with 2% hydrofluoric acid for about 10 min. With this treatment the glass phase is dissolved preferentially, while the mullite needles are more or less exposed. Therefore, they adhere to the first replica and remain there during further preparation so that they can be directly examined in the electron microscope—in contrast to the usual replicas which contain nothing from the original sample material. This is a special case of an *extraction replica* which will be dealt with in Section 2.3.3.

2.3.3. Extraction Replicas

A variation of the single-stage replica procedure consists in not only reproducing the surface profile with the film, but beyond that, in extracting certain minute constituents of the sample as, for example, crystalline precipitates with the film. This has the great advantage that these small particles can be directly investigated in transmission and also by electron diffraction (identification of unknown phases, inclusions, etc.). This method has been particularly useful for steels, since carbides, nitrides, and the like can be "extracted" while preserving their original distribution and arrangement within the sample.

The method consists of dissolving, with a suitable etchant, the matrix material of the sample, e.g., the ferrite matrix, faster than the inclusions, so that these protrude somewhat from the surface. The replica film is then

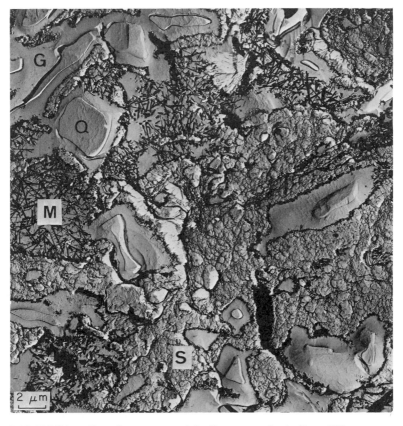

Fig. 2.26. Triafol replica of quartz porcelain (fracture surface). Four different structural components are distinguishable: Q = quartz, G = glass phase, S = primary mullite, M = secondary mullite (needle mullite, 4000×). (Courtesy Schüller [16].)

applied, usually by carbon evaporation or as a plastic film, Fig. 2.27A. With a second etching, the matrix material of the sample is further dissolved. This second etching, however, should not attack the particles to be extracted. If the latter are sufficiently "loosened," they will be lifted off with the carbon film into which they are embedded.

An example of an extraction replica from a Cr–Mo steel (10 CrMo 9 10) is given in Fig. 2.27B. The ferrite matrix was etched with a solution of 4 g picric acid and 1 cm³ hydrochloric acid 1.19 in ethyl alcohol. The extraction replica was made with Mowital F 40 dissolved in chloroform. This very thin replica film was strengthened by a double layer of collodion HP 5000 which had been dissolved in amyl acetate. After the composite replica was separated from the sample surface, the double layer of collodion was dissolved in amyl acetate without damage to the thin Mowital replica film.

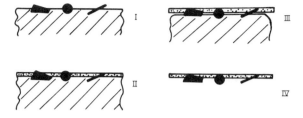

Fig. 2.27A. The four stages of extraction replication. I differential etching, II film application, III second (matrix) etching, IV lifting off of the extraction replica.

Fig. 2.27B. Extraction replica of a 10 CrMo 9 10 steel. Heat treatment: 20 min 930°C/air cooling plus 2 hr 680°C/air cooling. Ferrite with needle-shaped carbide precipitates of Mo_2C type. The larger, bulky particles are carbides of $M_{23}C_6$ type ($40,000 \times$). (Courtesy Schrader [17].)

The carbide particles which protrude from the ferrite matrix are wetted well by the Mowital solution and are firmly embedded by the drying plastic. When the double film is lifted from the sample surface, the small, usually brittle, carbide particles break off and remain embedded in the replica (according to Schrader [17]).

In conclusion, it should be emphasized again that almost all replica films are thinner and have poorer contrast than (e.g., electrolytically) thinned metals. While the latter, in general, are imaged with the highest available accelerating voltage, replicas are better examined with lower voltages (in the region from 50 to 80 kV). See Section 1.6.1 for a discussion of the voltage dependence of image contrast.

2.4. Preparation of Powders

Powders in general are not considered materials as such; however, in powder metallurgy, for example, they serve as raw material for sintered metallic composite materials. In addition, there are many other cases in which the electron microscopic investigation of powders, dusts, etc., is of interest, e.g., in pollution monitoring (important for ecology research).

First, a suitable *supporting film* is necessary on which to prepare the powder. The support film can be made from the same collodion solution which was described above for the preparation of plastic replicas. Because the support film should be thicker than a replica, a stronger concentration is recommended, e.g., 3–10% collodion solution in amyl acetate. Using a pipette, *one drop* only of this solution is placed on a water surface. The water is in a so-called suction filter (Fig. 2.28), onto whose sintered-glass filter screen the specimen grids to be coated have been placed. The drop

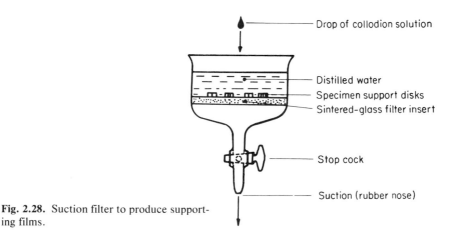

Drop of collodion solution

Distilled water
Specimen support disks
Sintered-glass filter insert

Stop cock

Suction (rubber nose)

Fig. 2.28. Suction filter to produce supporting films.

of plastic solution spreads over the water surface within 2–3 min to form a very thin film which can be recognized by its interference colors. The suction filter must stand absolutely motionless, since otherwise the film will wrinkle. The first plastic film created in this way is removed wtih a pair of tweezers and discarded—it serves only to clean the water surface of dust. The final film is produced with a second drop of plastic. This film is lowered onto the specimen grids by drawing off or draining the water. When the grids are dry, they are coated and ready to be used. One checks in the EM whether the plastic film is clean and without holes and tears. Another, easy way to produce supporting films is by evaporating carbon onto a clean glass surface (e.g., glass supports as used for light microscopy). The carbon film will float off easily from the smooth glass surface after inclined immersion into a water bath (cf. Section 2.3.1). (Before immersion the carbon film is scored with a pointed needle into small squares.) These film pieces are fished out from below with a grid or specimen aperture and dried carefully on filter paper.

The *powders* to be examined are prepared by making a suspension in water or another suitable suspension medium. The dilution must be chosen so that, on the one hand, the powder particles on the support film will not lie clustered on top of each other; and, on the other hand, a sufficient number of powder particles will be present in the field of view. A good dilution ratio usually lies between 1 : 100 and 1 : 10,000. For many applications, e.g., metal powder, distilled water is a good suspension medium. Kaolin powder (raw material for porcelain production) is suspended in an organic medium such as octane 80. Agglomeration of the powder in the suspension medium is prevented by a short ultrasonic treatment which, however, should not be too vigorous or too prolonged since otherwise there is danger of an additional unwanted breakup of the powder particles. The limits of beneficial ultrasonic treatment as well as the appropriate dilution of the suspension must be experimentally determined from case to case.

Now a drop of the suspension is placed onto one of the mounted supporting films. The suspension should form a hemispherical drop on the grid. (It must not run off!) Drying of the drop can be somewhat accelerated by mildly heating the specimen grid under an incandescent lamp. After the liquid is evaporated, the powder will adhere to the support film.

Figure 2.29 shows as an example a micrograph of Manganese nodule powder particles prepared on a carbon film. Manganese nodules occur on most deep-sea ocean floors and are considered to be essential resources for Cu, Ni, and Co in the future—they contain these elements as oxides and hydroxides together with Fe and Mn. For details of Mn nodule EM

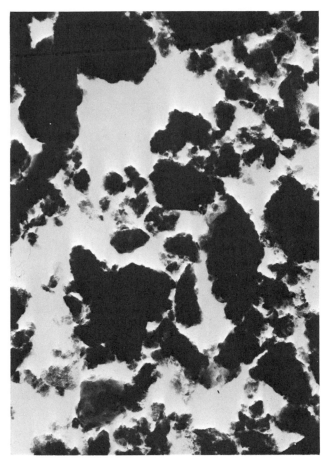

Fig. 2.29. Manganese nodule power DWHD-16 from v. Heimendahl *et al.* [32]. Metal analysis in wt %: 24.5 Mn, 11.5 Fe, 1.15 Co, 0.7 Ni, 0.25 Cu (40,000×).

preparation see von Heimendahl *et al.* [32], Fleischmann and von Heimendahl [33], and Gradel and von Heimendahl [34].

Appendix

The main suppliers of the items required for EM specimen preparation are

1) In the USA

    Ernest F. Fullam, Inc., PO Box 444, Schenectady, NY 12301;

Ladd Research Industries, Inc., PO Box 901, Burlington, VT 05401;
Pelco; Ted Pella, Inc., PO Box 510, Tustin, CA 92680;
2) In the United Kingdom
Aeon Laboratories, Beech Hill, Ridgemead Road, Englefield Green,
Egham, Surrey;
Polaron Equipment Ltd., 4 Shakespeare Road, London N3 1XH.

References

1. P. B. Hirsch, A. Howie, R. B Nicholson, D. W. Pashley, and M. J. Whelan, "Electron Microscopy of Thin Crystals." Butterworths, London, 1971.
2. G. Thomas and M. Goringe, "Transmission Electron Microscopy of Materials." Wiley, New York, 1979.
3. J. J. Theler, Métaux, Corrosion, Industrie, No. 496, p. 3 (1966).
4. M. von Heimendahl, *Prakt. Metallogr.* **4**, 65 (1967).
5. K. L. Maurer and H. Schäffer, *Prakt. Metallogr.* **4**, 388 (1967).
6. R. C. Glenn and W. R. Duff, *Trans. Q. ASM* **58**, 428 (1965).
7. E. Scheucher, *Pract. Metallogr.* **7**, 84 (1970).
8. L. Reimer, "Elektronenmikroskopische Untersuchungs- und Präparationsmethoden," 2nd ed., Springer, Berlin and New York 1967.
9. V. A. Phillips, *Prakt. Metallogr.* **4**, 637 (1967).
10. A. Krauth, *Glastechn. Ber.* **42**, 139 (1969).
11. M. H. Lewis, *Phil. Mag.* **14**, 1003 (1966); **17**, 481 (1968).
12. H. Alexander, *Phys. Status Solidi* **26**, 725 (1968).
13. H. Mayer, "Physik dünner Schichten," Bibliography, 2 Volumes. Wissenschaftliche Verlagsgesellschaft, Stuttgart, 1972.
14. K. H. Schüller, *Fortschr. Mineralog.* **45**, 281 (1968).
15. K. H. Schüller, *Ber. Deutsch. Keram. Ges.* **42**, 299 (1965).
16. K. H. Schüller, *Ber. Deutsch. Keram. Ges.* **40**, 320 (1963).
17. A. Schrader, *in* "De Ferri Metallographia," A. Schrader and A. Rose, eds., Vol. 2, Gefüge der Stähle pp. 127–130, Table 411, Verlag Stahleisen m.b.H., Düsseldorf, 1966.
18. B. Reppich, *Phys. Status Solidi* **35**, 339 (1969).
19. R. D. Schoone and E. A. Fischione, *Rev. Sci. Instrum.* **37**, 1351 (1966).
20. C. K. H. DuBose and J. O. Stiegler, *Rev. Sci. Instrum.* **38**, 694 (1967).
20a. M. von Heimendahl, W. Blum, R. Singer, and G. Schumann, Prakt. Metallogr. **15**, 269 (1978).
21. H. J. Rack and M. Cohen, *Metall. Trans.* **1**, 1050 (1970).
22. H. Bach, *J. Non-Cryst. Solids* **3**, 1 (1970).
23. D. J. Barber, *J. Mater. Sci.* **5**, 1 (1970).
24. W. J. McG. Tegart, "The Electrolytic and Chemical Polishing of Metals." Pergamon (Oxford), London and New York, 1959.
25. W. Machu, "Oberflächenbehandlung von Eisen und Nichteisenmetallen," 2nd ed., Leipzig, Geest and Portig Verlag, 1957.
26. P. Gillespie *et al.*, *J. Mater. Sci.* **6**, 87 (1971).
27. J. Petermann and H. Gleiter, *Phil. Mag* **25**, 813 (1972).
28. D. T. Grubb, A. Keller, and G. W. Groves, *J. Mater. Sci.* **7**, 131 (1972).
29. D. T. Grubb and G. W. Groves, *Phil. Mag.* **24**, 815 (1971).

30. M. von Heimendahl, *in* "Methodensammlung der Elektronenmikroskopie," (G. Schîm-
mel and W. Vogell, eds.), Wissenschaftl. Verlagsgesellschaft mbH., Stuttgart, 1972.

31. U. E. Wolff, *J. Nucl. Mater.* **40,** 230 (1971).

32. M. von Heimendahl, G. L. Hubred, D. W. Fuerstenau, and G. Thomas, *Deep-Sea Res.*
23, 69 (1976).

33. W. Fleischmann and M. von Heimendahl, *Mineral. Deposita* **12,** 155 (1977).

34. G. Gradel and M. von Heimendahl, *Mineral. Deposita* **14,** 219 (1979).

3. Electron Diffraction

3.1. Fundamentals, Comparison with X-Ray Diffraction

As explained in Chapter 1, electrons are diffracted by crystalline specimens; and here, as with x-rays, *Bragg's law* applies. In Section 1.6 image formation and the different types of contrast in transmission electron microscopy were treated, and the great importance of *diffraction* (in addition to *scattering*) was indicated, at least in qualitative terms. Before the quantitative details of *contrast theories* can be discussed (in the next chapter), the analysis of *electron diffraction patterns* has to be mastered.

The analysis of diffraction patterns involves so-to-speak two levels: (1) *geometry*, i.e., the position of interference lines or spots, and (2) *intensity*. From geometry alone a large amount of information can be gained. The present chapter is devoted to this part of the analysis. Chapter 4 deals with the contrast theories which are based on the evaluation of relative diffraction and/or scattering intensities.

In Table 3.1 the similarities and differences of x-ray and electron diffraction are juxtaposed.[1] Knowledge of x-ray structure analysis (including Miller indices) is a necessary prerequisite for students of this book [15]. With respect to electron diffraction some of the facts have been discussed in Chapter 1, some are self-explanatory, and others will become clear in the later part of this chapter.

As with x-ray diffraction, so with electron diffraction, *single-crystal patterns* and *polycrystalline diffraction patterns* are distinguished. Single crystals yield *spot patterns* (see Fig. 3.4, or, three dimensionally, Figs. 1.9 and 4.11). Polycrystalline materials yield *ring patterns* (Debye–Scherrer rings, see Fig. 3.3). Single-crystal patterns are produced when the electron beam illuminates but one individual crystallite. For most metal thin foils this condition is frequently met, especially for selected area diffraction (SAD, Section 1.6), since the SAD aperture limits the sampled area to a diameter of only 1–2 μm. If, however, the grain size is small and many grains are illuminated simultaneously, Debye–Scherrer rings are pro-

---

[1] A similar comparison which also includes neutron diffraction can be found in a monograph by Cohen [1].

Table 3.1 Comparison of X-Ray Diffraction (X) and Electron Diffraction (E)

Similarities	Differences
1. *Nature of superposition* of waves, leading to: *Bragg's law,* structure factor, extinction laws.	1. *Nature of scattering process* at individual atoms, i.e.: (E): scattering by atom nucleus, (X): scattering by shell electrons.
2. *Types of patterns:* Single-crystal (Laue), Debye–Scherrer, texture patterns.	2. *Wavelengths of radiation:* (E): So small that diffraction angle is only 0–$2°$, therefore $\sin \theta \approx \theta$; diffraction pattern is approx. a plane section through reciprocal lattice. (X): All diffraction angles up to $180°$ occur; therefore the locus of the diffraction spots is the Ewald *sphere.*
3. Possibility of determining *crystal orientation* from single-crystal patterns, (however, compare point 7 under differences).	3. *Intensity of diffraction spots:* Because stronger interaction with the atom nucleus occurs with (E), intensity is 10^6–10^7 times that observed with (X).
	4. *Penetration of radiation:* As a consequence of point 3, with (E) order of magnitude 1 μm or less, with (X) order of magnitude 100 μm.
	5. *Affected sample volume:* As a consequence of point 4 and of the beam cross section, for (E) approx. 1 μm^3 $= 10^{-9}$ mm^3, for (X) 0.1 to 5 mm^3.
	6. *Accuracy of crystallographic orientation determination:* for (X) $1°$ or better, for (E) $\pm 5°$ or worse if only spot patterns are used; better if Kikuchi lines, center-of-gravity methods, or similar special procedures are used.
	7. *Unambiguity of orientation determination:* With (X) the Laue pattern generally has a onefold symmetry; consequently the analysis is *unambiguous.* The diffraction pattern with (E) has a twofold symmetry; consequently the analysis of a *single diffraction pattern* generally is ambiguous. (Only those orientations are indistinguishable which are produced by a rotation of $180°$ around the primary beam.)

duced. These are continuous only if the grain size is sufficiently small ($\ll 1$ μm). Otherwise the rings are spotty.

By analysis of either type of pattern the lattice parameters and/or lattice spacings (*d*-values) of the *diffracting sample* area can be *determined*. This is an important benefit of electron diffraction in many metallurgical investigations. Besides, the crystallographic orientation of the crystal can be determined from single-crystal patterns. A very important practical application is the solution of the problem how to determine the crystallographic *directions and planes* in a given TEM image (so-called trace analysis, Section 3.8).

Patterns of polycrystals are the subject of Section 3.2. Sections 3.3–3.11 deal with the analysis of single-crystal patterns. The remainder of Section 3.1 applies equally to both kinds of diffraction patterns.

One of the most important characteristics of electron diffraction is the *very small diffraction angle* which is a consequence of the small wavelength λ of electrons. The calculation given in Section 1.6 showed that, e.g., the diffraction angle 2θ for the (111) reflection of aluminum with 100 kV electrons is only 0.92°. For higher index reflections the resulting diffraction angles 2θ are 1–3°. The angle θ between the reflecting lattice planes and the primary beam, therefore, is between 0 and 1.5°. This fact leads to the important:

Theorem 1. *The reflecting lattice planes are nearly parallel to the primary beam.*

In crystallography those planes which contain a common direction are also called the "planes of a zone." The common direction is the "zone axis" (Fig. 3.1). Theorem 1 can now be reformulated:

Theorem 1. (alternative version). *Reflections are obtained from only those lattice planes which have the primary beam as a zone axis.*

Because of the small diffraction angle, the function $\sin \theta$ in the Bragg equation can be replaced with θ. This simplification has an important consequence. Figure 3.2 shows a longitudinal section containing the primary beam, the diffracted beam, the diffracting sample, the image screen (or photographic plate), and, their sizes greatly exaggerated, the Bragg angle θ and the diffracting angle 2θ. L is the distance between sample and photographic plate, the so-called *camera length,* and R the distance on the plate between the transmitted beam (zero order) and a diffraction spot.

Figure 3.2 shows that $\tan \theta = R/2L$; and because of the small value of θ it follows that $\theta = R/2L$. Consequently, the Bragg equation can be

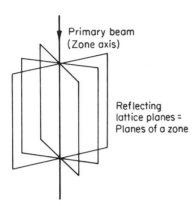

Fig. 3.1. Primary beam as zone axis.

reformulated:

$$n\lambda = 2d \sin \theta = 2d\theta = 2dR/2L.$$

Since in this book n is always set as 1 (i.e., higher orders are taken care of by corresponding multiples of the Miller indices), the above equation leads to the

Basic Equation $\boxed{\lambda L = Rd}$.   (3.1)

For the monoenergetic electrons used in an EM, the product λL is a constant. It is called the *diffraction constant* or *camera constant*.

For most of the commonly used 100 kV electron microscopes the camera length L is about 500 mm. With a wavelength $\lambda = 0.0037$ nm (100 kV) the camera constant λL is approximately

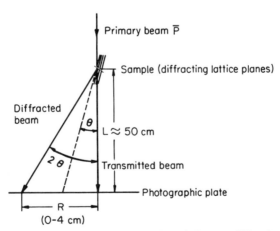

Fig. 3.2. Diagram for derivation of Basic Equation of electron diffraction (simplified ray path; the effect of focusing in the EM is omitted).

1.8–2 [mm × nm]. When electron diffraction patterns are to be analyzed this value has to be known accurately, i.e., it has to be calibrated with standards (Section 3.2) or by other procedures to be discussed. The camera constant is not an instrument constant because L, as well as R, is subject to small variations. These are caused by differences in the exact position of the sample and by variations in the nominal magnifications set by the projector lens. Frequent standardization of the camera constant provides, as an additional benefit, a check of the actual wavelength. (Deviations from the nominal value of the high voltage will be noticed, indicating malfunctioning of the microscope).

3.2. Debye–Scherrer Patterns, Standardization

The analysis of ring patterns (as in Fig. 3.3) is very simple. If the crystal structure and the lattice constant(s) of the diffracting sample are known the first (largest) interplanar spacings d can be calculated. Substitution of these d-values $d_1, d_2, d_3, \ldots$ into the Basic Equation [Eq. (3.1)] yields the ring radii $R_1, R_2, R_3, \ldots$ to be expected if the camera constant λL is known. Conversely, from the measured ring radii and the known d-values the camera constant can be determined independently for each ring. In this case the sample has been used as a *standard* for calibrating λL.

The ring pattern of an unknown material can be analyzed, if λL is known, by determining the interplanar spacings d_n from the measured ring radii R_n. With this set of d-values the unknown material can often be identified with the help of the ASTM Card File [2] which contains over 10,000 crystalline substances arranged according to d-values. For such a search, a rough estimate of the ring intensities (strong–medium–weak) is also useful.

If the rules [15] known from x-ray diffraction are applied to the sequence of ring radii R_n, the unknown material can often be quickly identified: For example, according to the extinction laws, face-centered cubic (fcc) lattices produce reflections only with those interplanar spacings d for which the Miller indices (hkl) are unmixed (i.e., either all odd or all even). From Eqs. (1.7) and (3.1) it follows that R is proportional to $(h^2 + k^2 + l^2)^{1/2}$. Consequently, the radii of the first rings of fcc crystals have to have the ratios $\sqrt{3}:\sqrt{4}:\sqrt{8}:\sqrt{11}$, etc. For body-centered cubic (bcc) crystals the extinction laws state that the sum $(h + k + l)$ has to be even, i.e., the ring radii of bcc crystals have the ratios $\sqrt{2}:\sqrt{4}:\sqrt{6}:\sqrt{8}$, etc. Thus these sequences serve to distinguish fcc and bcc crystals. Similar rules apply to other crystal systems.

Few *materials* are so fine grained that they give rise to ring patterns in selected area diffraction. Ring patterns, however, are important for EM calibration. Therefore, fine-grained *standards* are used to determine the camera constant λL of a microscope under given experimental conditions. Good standards are substances which are chemically stable, unaffected by the electron beam, and easy to prepare for the EM. Three

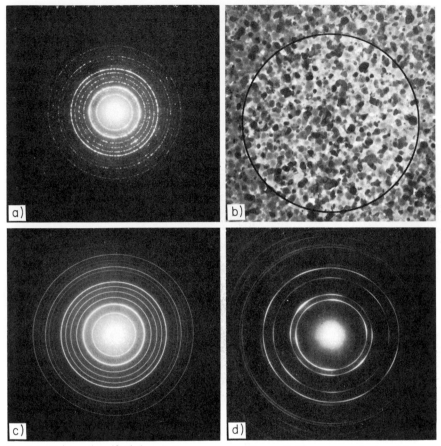

Fig. 3.3. Comparison of Debye–Scherrer patterns from a CsCl structure a)–c) and from a fcc structure d). a)–c) Calibration standard TlCl, evaporated onto a Mowital support film. a) Selected area diffraction (SAD) of 2.35 μm diameter area circled in b) (selected area aperture 50 μm diameter, 100 kV). b) Image of same area, 20,000×. c) Diffraction of same sample without imaging lenses, area ~100 μm diameter, 100 kV. d) Diffraction pattern of gold leaf taken without lenses (same sample as in Fig. 2.9). The intensity variations around the circumference of the rings are due to the deformation texture (from hammering the leaf). Conditions of photograph as in c).

substances have been proven particularly suitable: thallium chloride TlCl (CsCl type, $a = 0.3841$ nm), lithium fluoride LiF (NaCl type, $a = 0.4020$ nm), and MgO (NaCl type, $a = 0.4202$ nm)[3]. TlCl and LiF can be evaporated from a tungsten boat. MgO can be deposited from the smoke which forms when a piece of Mg strip or wire is burned. Also evaporated pure gold or aluminum are approved as standard materials.

The calibration standard is best deposited directly onto the sample for which an exact value of λL is needed.[2] If both patterns have few and non-overlapping lines, or one is a spot pattern and the other a ring pattern, the analysis of the superimposed patterns does not pose a problem. If the calibration pattern and the sample pattern have to be taken one after the other, there is danger that the (height) position of the two objects in the EM is not identical. As a consequence, the magnifications may be different, resulting in different values for λL. Control of identical conditions for such sequential photographs is therefore imperative.

Figure 3.3 shows patterns of the calibration standard TlCl as an example. Figure 3.3a is a selected area diffraction (SAD) pattern of that area of the TlCl layer circled in Fig. 3.3b (2.35 μm diameter). With the SAD technique discussed in Section 1.6.2, the diffraction pattern of only this small area is produced in the back focal plane and projected onto the image screen by the subsequent lenses. However, *diffraction* in the EM can also be accomplished *without imaging lenses:* For this method, only the two condenser lenses are activated and the cross-over is focused on the image screen. If a diffracting sample is in the beam the diffracted rays are also focused. In this case, however, a much larger area of about 100 μm diameter contributes to the diffraction pattern (Fig. 3.3c). For this reason diffraction rings have a uniform intensity, while with SAD comparatively few crystallites contribute to the pattern which consequently is "grainy," i.e., the rings consist of individual spots. Because of the different optical paths in the two cases, the effective camera constant λL is different. During SAD one has to work with a strongly defocused beam because TlCl tends to decompose even under moderate electron irradiation.

Table 3.2 gives an application example for the determination of λL. The interplanar spacings d with the listed (hkl) values have been determined

Table 3.2 Analysis of SAD Debye–Scherrer Pattern[a]

hkl	$h^2 + k^2 + l^2$	d (nm)	R (mm)	λL^b
100	1	0.384	4.6	1.766
110	2	0.272	6.6	1.795
111	3	0.222	8.0	1.776
200	4	0.192	9.3	1.785
210	5	0.172	10.5	1.806
211	6	0.157	11.5	1.805

[a] Fig. 3.3a.
[b] Mean camera constant $\lambda L = 1.789$ [mm × nm].

---

[2] It is recommended to cover part of the sample during the deposition of the standard (e.g., with a grid). This way areas with and without the standard can be examined.

for TlCl (a = 0.384 nm). From the measured ring radii (Fig. 3.3a) the product $Rd = \lambda L$ has been calculated.

For the diffraction pattern without lenses, Fig. 3.3c, an analogous value of λL = 2.20[nm × mm] is found. For precise measurements of rings or diffraction spots precision glass or steel rulers are recommended coupled with magnifiers and light boxes, or one of the well-known measuring instruments used for x-ray diffraction analysis.

All further diffraction patterns reproduced in this book are selected area diffraction (SAD) patterns.

3.3. Reciprocal Lattice

The Basic Equation (3.1) can also be expressed as

$$R = \text{const}/d. \tag{3.2}$$

As a consequence the following theorems hold:

a) The distance R of each diffraction spot from the primary beam, the value of its *position vector* **R**, is inversely proportional to the interplanar spacing d, and

b) **R** is normal to the diffracting lattice planes if these are thought of as being located at the center of the diffraction pattern (Consequence of Theorem 1).

These last two theorems, except for a dimensional factor, see below, are also the definitions of a point in the so-called "reciprocal lattice." The latter is defined as follows: The *real lattice* built of lattice atoms and with lattice planes (*hkl*) of interplanar spacing d (*hkl*) is thought of as located at the origin of a coordinate system. With this real lattice a system of points, the *reciprocal lattice,* is unambiguously associated in the following way: Each *family of lattice planes* (*hkl*) is associated with a *point* g, also identified by (*hkl*), whose distance from the origin of the coordinate system equals 1/d and which is located on the normal to the lattice planes (*hkl*) (Fig. 3.4). There is a direct correspondence between the total of all lattice planes of the real lattice and the total of all points of the reciprocal lattice. From this definition and the two theorems a) and b) above the important Theorem 2 can be formulated:

Theorem 2. *Each diffraction spot represents the reciprocal lattice point of the associated family of lattice planes (hkl). The total diffraction pattern is a scaled image of a nearly plane section through the reciprocal lattice, normal to the primary beam **P**.*

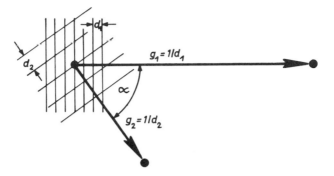

Fig. 3.4. Schematic to explain definition of Reciprocal Lattice. The center of the real lattice is the origin of the reciprocal lattice. Each family of lattice planes with interplanar distance d is associated with a reciprocal lattice point g (vector $\mathbf{g}$) at a distance $1/d$ and in a direction normal to the lattice planes.

This theorem is only true as a first approximation because the Bragg angles θ, although small ($<1.5°$), are not zero. The consequences of these small deviations will be discussed later in some detail.

It should be mentioned here, however, that the Bragg reflection condition (angle of incidence $= \theta$) is not a very rigorous condition (compare Section 4.3). Consequently, reflections are, in fact, obtained frequently from many of the planes of one zone (mentioned in Theorem 1). Therefore a whole spot pattern is generated, not only a single diffraction spot. The latter would be expected for a given d and θ and for a monoenergetic electron beam of a given wavelength λ, if the Bragg equation (1.1) were strictly satisfied.

Just as the positions of the atoms form a three-dimensional array in the real lattice, so do the points of the reciprocal lattice. Both are *translation lattices* which can be thought of as generated by translation, or repeated placement side by side, of a *unit cell*. In the *real lattice* the unit cell consists of the unit vectors (lattice parameters) $\mathbf{a}$, $\mathbf{b}$, and $\mathbf{c}$, from which each lattice point $\mathbf{r}$ is obtained by

$$\mathbf{r} = m\mathbf{a} + n\mathbf{b} + p\mathbf{c} \quad [\text{nm}], \tag{3.3}$$

where m, n, and p can be any integers.

The *unit cell* of the *reciprocal lattice* consists of the unit vectors (translation vectors) $\mathbf{a}^*$, $\mathbf{b}^*$, and $\mathbf{c}^*$. The lattice points of the reciprocal lattice are customarily called $\mathbf{g}$ and can be described by:

$$\mathbf{g}(hkl) = h\mathbf{a}^* + k\mathbf{b}^* + l\mathbf{c}^* \quad [\text{nm}^{-1}]. \tag{3.4}$$

The asterisk is a reminder that the unit vectors belong to the reciprocal lattice and have the dimension nm^{-1}.

According to the above definition

$$|\mathbf{g}|(hkl) = g(hkl) = \frac{1}{d(hkl)} \quad [\text{nm}^{-1}].$$

The *model of a reciprocal lattice* will be found to be a very useful tool which will facilitate many thought processes and derivations.

From here on one must always distinguish between reciprocal space (dimension nanometer^{-1}) and real space (nanometer). According to Theorem 2 the diffraction pattern with its vectors $\mathbf{R}_n(hkl)$ defining each diffraction spot is a true-to-scale image of the reciprocal lattice. The link between the two lattices, or the "scale factor," between a distance $|\mathbf{R}| = R$ in millimeter on the photographic plate and the analogous distance $|\mathbf{g}| = g$ in nanometer^{-1} in reciprocal space is given by the definition $g = 1/d$ and the Basic Equation as follows:

$$\boxed{R\ [\text{mm}] = \lambda L \cdot g \quad [\text{nm}^{-1}]}\ .$$

Thus the scale factor is the camera constant λL. If it is, for example, 2 [mm × nm], then 10 mm on the photographic plate is equivalent to 5 nm^{-1} in reciprocal space.

Given the definition of the reciprocal lattice one can also prove that the unit vectors $\mathbf{a}^*$, $\mathbf{b}^*$, and $\mathbf{c}^*$ have the following relation to $\mathbf{a}$, $\mathbf{b}$, and $\mathbf{c}$:

$$\mathbf{a}^*\,\mathbf{a} = \mathbf{b}^*\,\mathbf{b} = \mathbf{c}^*\,\mathbf{c} = 1 \quad \text{and} \quad \mathbf{a}^*\,\mathbf{b} = \mathbf{b}^*\,\mathbf{c} = \mathbf{a}^*\,\mathbf{c} = \cdots = 0. \quad (3.5)$$

These equations, too, can serve to define the reciprocal lattice. They are true for arbitrary lattices, also those with oblique angles. The vector $\mathbf{a}^*$ is normal to $\mathbf{b}$ and $\mathbf{c}$, i.e., normal to the plane containing $\mathbf{b}$ and $\mathbf{c}$. Likewise, $\mathbf{b}^*$ is normal to $\mathbf{a}$ and $\mathbf{c}$, and $\mathbf{c}^*$ is normal to $\mathbf{a}$ and $\mathbf{b}$. If $\mathbf{a}$ is *not* normal to $\mathbf{b}$ and $\mathbf{c}$, then $\mathbf{a}^*$ is *not* parallel to $\mathbf{a}$. For *orthogonal* lattices, however,

$$\mathbf{a}^*\|\mathbf{a},\ \mathbf{b}^*\|\mathbf{b},\ \mathbf{c}^*\|\mathbf{c} \quad \text{and} \quad |\mathbf{a}^*| = 1/|\mathbf{a}|,\ |\mathbf{b}^*| = 1/|\mathbf{b}|,\ |\mathbf{c}^*| = 1/|\mathbf{c}|.$$

Note that the definition equations (3.5) are *symmetric* with respect to $\mathbf{a}^*$ and $\mathbf{a}$, etc. Therefore the following theorem holds: If R is the reciprocal lattice of the real lattice L, then conversely L is the reciprocal lattice for R.

Furthermore, it can be shown that

$$\mathbf{a}^* = \frac{\mathbf{b} \times \mathbf{c}}{V},\ \mathbf{b}^* = \frac{\mathbf{c} \times \mathbf{a}}{V},\ \mathbf{c}^* = \frac{\mathbf{a} \times \mathbf{b}}{V},\ \text{with}$$

$$V = \mathbf{a}\,(\mathbf{b} \times \mathbf{c}) = \mathbf{b}\,(\mathbf{c} \times \mathbf{a}) = \mathbf{c}\,(\mathbf{a} \times \mathbf{b}).$$

V is the unit cell volume of the real lattice L.

3.4. Construction of Simple Diffraction Patterns

With the help of Theorems 1 and 2 it is possible, in principle, to construct the diffraction patterns to be expected. One pretends to look from above—along the primary beam **P**—onto the crystal and considers which planes with what d-values are parallel to **P**. If, e.g., **P** is parallel to a cube edge [001] of a fcc lattice the diffraction spots drawn in Fig. 3.5 are generated. One has to take into account that the same extinction laws operate as in x-ray diffraction, i.e., reflections occur only on planes with unmixed indices. Each diffraction spot hkl obeys Theorems 1 and 2 as well as the Basic Equation (3.1). If eq. (1.7) is substituted into eq. (3.1), the distance R of the diffraction spot from the origin is

$$R(hkl) = \lambda L \, (h^2 + k^2 + l^2)^{1/2}/a.$$

For example, for $\lambda L = 1.9$ [mm × nm] and aluminum ($a = 0.405$ nm) the distance of the {200} reflections is $R\{200\} = 9.38$ mm. The *rules of vector addition* [Eq. (3.4)] hold for the vectors $\mathbf{R}(hkl)$ which lead to the hkl diffraction spots. The crystal with all its lattice planes of type $(hk0)$ is sketched in the origin (000) of the diffraction pattern, which, of course, could extend in all directions within the plane of the paper. The reciprocal lattice points (100) and (010) at the end of the unit vectors $\mathbf{a}^*$ and $\mathbf{b}^*$ in Fig. 3.5 are not capable of reflection because of the extinction laws. When diffraction spots are indexed the parantheses are often omitted for simplicity's sake.

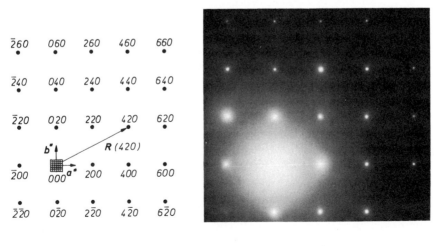

a)   b)

Fig. 3.5. Diffraction pattern of a fcc lattice in [001] orientation. a) Schematic and b) diffraction pattern of an Al foil.

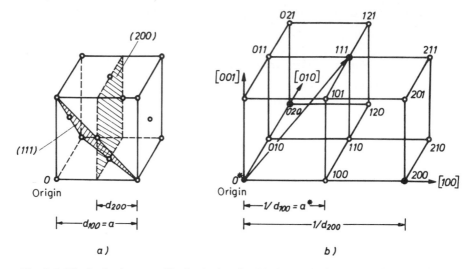

Fig. 3.6. The fcc lattice: a) real lattice (unit cell) with the two lattice planes of lowest indices (111) and (200). b) Part of the associated reciprocal lattice. Only the filled-in black lattice points {111} and {200} correspond to planes capable of reflection, and only these appear in the diffraction pattern. (After Thomas [1a].)

The diffraction pattern represents a cut through the reciprocal lattice along an (001) plane. The diffracting crystal is in the form of a thin transpararent foil. Thus, the statement that the primary beam is parallel to the [001] direction means that this is also the direction of the foil normal. For this statement to be true the foil has to be exactly normal to **P**; but this is not necessarily the case.[3] However, the usage "[hkl] orientation parallel to foil normal" has generally been adopted, although the more exact statement should be "[hkl] direction parallel to the primary beam **P**."

It would be very cumbersome to construct, in the manner described all possible diffraction patterns which can be generated in space by the different lattice types and in different orientations. The goal to index all experimentally observed diffraction patterns can better be reached in three steps with each step a more generalized concept.

First, Theorem 3 describes the conditions for *cubic lattices*.

Theorem 3. The reciprocal lattice of a primitive cubic real lattice with lattice constant a is also primitive cubic and has the lattice constant $a*$

---

[3] A procedure for the analytical treatment of the possible difference between **P** and the foil normal is given by Schwartzkopff [4].

$= 1/a$; the reciprocal lattice of a fcc real lattice with lattice constant a is a bcc lattice with lattice constant $a^* = 2/a$; the reciprocal lattice of a bcc real lattice (a) is a fcc lattice with lattice constant $a^* = 2/a$. In all three cases the corresponding axes of the cubic real lattice and the cubic reciprocal lattice are parallel.

The proof of Theorem 3 is so simple that the reader can find it himself by stepwise construction of the rel-points[4] using Fig. 3.6 and applying the extinction laws.

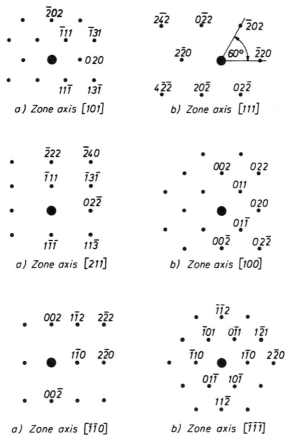

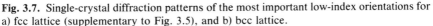

Fig. 3.7. Single-crystal diffraction patterns of the most important low-index orientations for a) fcc lattice (supplementary to Fig. 3.5), and b) bcc lattice.

[4] Points, directions, and planes in the reciprocal lattice are characterized by the abbreviated prefix "rel."

Theorem 3 facilitates the construction and interpretation of a number of simple diffraction patterns of fcc and bcc real lattices which have practical importance, in particular those in (100), (110), and (111) orientation. With the help of three-dimensional models of the relevant reciprocal lattices one can easily visualize how the respective sections normal to **P** through the reciprocal lattice would look. If Figs. 3.5 and 3.6 are used as examples and Eq. (3.4) is applied, the *indexing* of the rel-points is also easy. The (001) section of Fig. 3.5 corresponds to the plane normal to the paper in Fig. 3.6. Figure 3.7 shows the appearance of a few simple diffraction patterns of fcc and bcc real lattices obtained by this method of construction.

The literature [5] contains many schematic pictures of the most important diffraction patterns of the most common crystal lattice types. Eichen *et al.* [6] have published a sort of "atlas" of diffraction patterns of fcc, bcc, hexagonal close-packed (hcp), and diamond cubic lattices. However, the analyst should be warned not to rely solely on comparison of diffraction patterns with those in an "atlas." It is indispensible that he master the art of systematic analysis of all possible diffraction patterns, for example, with the methods described here.

Theorem 2 is valid for all lattice types, also noncubic ones. With the latter, however, it is difficult to visualize, in detail, the diffraction pattern on the basis of sections through the reciprocal lattice.[5] In these cases, the general method of indexing described in Section 3.6 will usually be used. But even for cubic crystals, the method using sections through the reciprocal lattice becomes rather involved for higher-indexed orientations as, e.g., (513). Here the second of the above-mentioned three steps will help, namely the "method of R_n ratios" (after Thomas).

3.5. Method of R_n Ratios

This method can be used *only for cubic lattices*, but for *these* the *general case* can be solved, i.e., diffraction patterns from arbitrary crystal orientations can be indexed.

Even in the general case the diffraction pattern has twofold symmetry. Rotation by 180° around **P** will result in the identical pattern, as is also seen in Figs. 3.5 and 3.7. This is so by reason of Theorem 1 and because the diffraction pattern is a plane section through the reciprocal lattice. Each diffraction pattern is formed by vector addition of only two vectors, in the following termed "basic vectors" (of the particular diffraction pat-

---

[5] The reciprocal lattices of noncubic real lattices are described in the text books on crystallography and x-ray diffraction.

tern).[6] In contrast to the special cases of Figs. 3.5 and 3.7, however, the two basic vectors in the general case may have different lengths and a priori may enclose an arbitrary angle, as, for example, the two basic vectors R_1 and R_2 in the diagram of Fig. 3.8. The diagram is representative of an arbitrary diffraction pattern. Specifically, this is an fcc diffraction pattern and is to be analyzed. The following values were measured:

$$R_1 = |R_1| = 1.92 \text{ cm}, R_2 = 3.25 \text{ cm}, \quad \text{and} \quad R_3 = R_1 + R_2,$$
$$\text{with} \quad R_3 = 4.25 \text{ cm}, \text{ angle } (R_1 R_2) = \alpha = 72.5°.$$

The method is based on the fact that for cubic lattices, as a consequence of Eq. 1.7 and the Basic Equation (3.1), it follows:

$$R = \frac{\lambda L}{a} (h^2 + k^2 + l^2)^{1/2} = \text{const} \cdot (h^2 + k^2 + l^2)^{1/2}, \qquad (3.6)$$

i.e., all experimentally measured R values have to be proportional to the square roots of the sum of the squares of the indices of the reflecting lattice planes. Consequently,

$$\frac{R_1}{R_2} = \frac{(h_1^2 + k_1^2 + l_1^2)^{1/2}}{(h_2^2 + k_2^2 + l_2^2)^{1/2}}. \qquad (3.7)$$

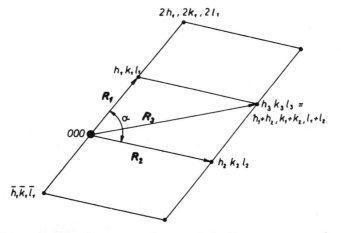

Fig. 3.8. Schematic diffraction pattern of a crystal of arbitrary structure and orientation (twofold symmetry).

[6] The basic vectors should not be confused with the unit-cell vectors of the reciprocal lattice. Only by chance could they be identical; in general the basic vectors are obtained from the unit-cell vectors according to (3.4).

It is useful to tabulate the possible ratios of the square roots in Eq. (3.7) for the fcc and bcc lattices or any cubic lattices with known extinction laws. Table 3.3 is valid for the fcc and the diamond cubic lattices, Table 3.4 for the bcc lattices. On a diffraction pattern photograph, R_1/R_2, R_2/R_3, etc., are measured as accurately as possible, and the table is searched for ratios of the same magnitude (considering the error limits). From the table the Miller indices h_n, k_n, l_n, associated with R_n, can be read directly, if the association was unambiguous and the measurements were sufficiently accurate.[7] To find the right combination, in practice a choice often has to be made, because of measuring inaccuracies, between several possible square root ratios. In the numerical example above $R_2 : R_1$ = 3.25 : 1.92 = 1.69. This ratio is close to the table value of 1.66 for the square root ratio [Eq. (3.7)] from the reflections {311} and {200}. Next, $R_3 : R_1$ = 4.25 : 1.92 = 2.21. This value is ambiguous since it lies between the theoretical possibilities of 2.18 (for {331} and {200}) and 2.235 (for {420} and {200}). A decision is possible by *analyzing the angle* between

Table 3.3 Ratios of Interplanar Spacings according to Eq. (3.7) for fcc Lattices and Diamond Cubic Lattices[a,b]

hkl	111	200	220	311	331	420	422	511 333	531
111	1^b								
200	1.155	1							
220	1.63^b	1.414	1^b						
311	1.92^b	1.66	1.17^b	1^b					
222	2.00	1.73	1.225	1.045					
400	2.31^b	2.00	1.415^b	1.21^b					
331	2.52^b	2.18	1.54^b	1.31^b	1^b				
420	2.58	2.235	1.58	1.35	1.027	1			
422	2.85^b	2.45	1.73^b	1.48^b	1.124^b	1.096	1^b		
333,511	3.00^b	2.60	1.84^b	1.57^b	1.19^b	1.16	1.06^b	1^b	
440	3.27^b	2.83	2.00^b	1.71^b	1.30^b	1.217	1.156^b	1.09^b	
531	3.42^b	2.96	2.09^b	1.785^b	1.36^b	1.32	1.21^b	1.14^b	1^b
600,442	3.46	3.00	2.12	1.81	1.38	1.34	1.225	1.157	1.014
620	3.66^b	3.16	2.24^b	1.91^b	1.45^b	1.42	1.29^b	1.22^b	1.07^b
533	3.79^b	3.28	2.32^b	1.98^b	1.503^b	1.47	1.34^b	1.26^b	1.11^b
622	3.82	3.32	2.34	2.00	1.52	1.48	1.355	1.28	1.12
444	4.00^b	3.47	2.45^b	2.09^b	1.59^b	1.55	1.415^b	1.33^b	1.17^b
711,551	4.12^b	3.57	2.52^b	2.15^b	1.64^b	1.595	1.458^b	1.374^b	1.207^b

[a] After Nolder and Thomas [7].
[b] Values applicable to the diamond cubic lattice.

[7] Following an idea by Roser [8], a special slide rule can be made by pasting onto the free back of a regular slide rule a special scale ruled according to Table 3.3 or Table 3.4. The analysis of diffraction patterns is hereby greatly accelerated.

Table 3.4 Ratios of Interplanar Spacings According to Eq. (3.7) for bcc Lattices[a]

hkl	110	200	211	310	222	321	411	420	332	510 431	521	530 433
110	1											
200	1.415	1										
211	1.73	1.225	1									
220	2.00	1.415	1.155									
310	2.235	1.58	1.29	1								
222	2.45	1.73	1.415	1.095	1							
321	2.645	1.87	1.53	1.185	1.08	1						
400	2.83	2.00	1.63	1.265	1.155	1.07						
411,330	3.00	2.12	1.73	1.34	1.225	1.135	1					
420	3.16	2.235	1.825	1.415	1.29	1.195	1.055	1				
332	3.315	2.345	1.915	1.485	1.355	1.255	1.105	1.05	1			
422	3.465	2.45	2.00	1.55	1.415	1.31	1.155	1.095	1.045			
510,431	3.605	2.55	2.08	1.61	1.47	1.365	1.20	1.14	1.09	1		
521	3.875	2.74	2.235	1.73	1.58	1.465	1.29	1.245	1.17	1.075	1	
440	4.00	2.83	2.31	1.79	1.63	1.51	1.335	1.265	1.21	1.11	1.035	
530,433	4.125	2.915	2.38	1.845	1.685	1.56	1.375	1.305	1.245	1.145	1.065	1
600,442	4.245	3.00	2.45	1.895	1.73	1.605	1.415	1.34	1.28	1.18	1.095	1.03
611,532	4.36	3.08	2.52	1.95	1.78	1.65	1.455	1.38	1.315	1.21	1.125	1.06
620	4.47	3.16	2.58	2.00	1.825	1.69	1.49	1.415	1.35	1.24	1.155	1.085
541	4.585	3.24	2.645	2.05	1.87	1.73	1.53	1.45	1.38.	1.27	1.185	1.11
622	4.69	3.315	2.71	2.10	1.915	1.77	1.565	1.485	1.415	1.30	1.21	1.135
631	4.795	3.39	2.77	2.145	1.955	1.815	1.60	1.515	1.445	1.33	1.24	1.16
444	4.90	3.465	2.83	2.19	2.00	1.85	1.635	1.55	1.48	1.36	1.265	1.185
710,550,543	5.00	3.535	2.89	2.235	2.04	1.89	1.665	1.58	1.51	1.385	1.29	1.21
640	5.10	3.605	2.94	2.28	2.08	1.925	1.70	1.61	1.54	1.415	1.315	1.235
721,633,552	5.195	3.675	3.00	2.325	2.12	1.955	1.73	1.645	1.57	1.44	1.34	1.26
642	5.29	3.74	3.055	2.365	2.16	2.00	1.765	1.675	1.595	1.47	1.365	1.285
730	5.385	3.81	3.11	2.41	2.20	2.035	1.795	1.705	1.625	1.495	1.39	1.305

[a] After Nolder and Thomas [7].

111

the R_n. Since the R_n are normal to the associated reflecting lattice planes (hkl), the angle between two vectors R_n is identical with the angle between the associated lattice planes. In cubic systems, the cosine of the angle α between the lattice planes (h_1, k_1, l_1) and (h_2, k_2, l_2) (see also Fig. 3.4) is

$$\cos \alpha = \frac{h_1 h_2 + k_1 k_2 + l_1 l_2}{[(h_1^2 + k_1^2 + l_1^2)(h_2^2 + k_2^2 + l_2^2)]^{1/2}} \tag{3.8}$$

The tentatively assumed indices are now tested by comparing the measured angles with those obtained with these indices. A check of the angles with Eq. (3.8) is necessary also for another reason: The ratio method yields from the table only the *type* $\{hkl\}$ of indices for the respective reflections; it does not give the complete indices (hkl), including sign and sequence of the individual indices. The latter can easily be found from Eq. (3.8) by trial and error. In this manner it is quickly determined that the correct indexing for the example of Fig. 3.8 is $(h_1 k_1 l_1) = (\bar{2}00)$, $(h_2 k_2 l_2) = (\bar{1}13)$, and $(h_3 k_3 l_3) = (\bar{3}13)$. With these indices the rule of vector addition $R_1 + R_2 = R_3$ is also satisfied.

In principle, there are always several (maximally 24) correct ways of indexing a given pattern. This is due to the fact that the choice how to designate the three axes (100), (010), and (001) in a real cubic lattice is arbitrary. Likewise, fixing the sequence and signs of the indices h, k, and l for the *first* indexed reflection is arbitrary. After that, indexing of the other reflections has to be "internally consistent," i.e., Eqs. (3.7) and (3.8) and the vector addition rules have to be satisfied for *all reflections*.

It should be emphasized once more that this method, which is universally applicable to all cubic lattices, has the great advantage that neither the interplanar spacings $d(hkl)$ nor the camera constant λL of the microscope need be known. However, after completed indexing it is recommended to make a check by calculating the product $Rd = \lambda L$ using the now known d-values. This product should have the same value for all reflections. This calculated value will show, on the one hand, the magnitude of the experimental error. On the other hand, it is good to know and to check the camera constant λL. Given the deviations caused by fine focusing and inaccuracies of sample position, the camera constant should change only within certain small limits, even during extended periods (compare Section 3.1) (practical example: long-time variation 1.85 ± 0.10 [mm × nm], measuring error ± 0.02 [mm × nm]). Besides, the knowledge of λL is necessary for the general case of indexing diffraction patterns of noncubic lattices, which will be discussed next.

3.6. General Case of Indexing Single-Crystal Diffraction Patterns

Consider a spot diagram of a crystal with arbitrary crystal structure and arbitrary orientation, as shown schematically in Fig. 3.8. If the crystal structure is completely unknown, the analysis is extremely difficult, since in most cases one is reduced to a trial and error method. Assume, therefore, that either the crystal structure is known, or, as is frequently the case in practice, there is a choice between only a few possible structures.

Indexing or identification, and the determination of the crystal orientation, can be done with the aid of the seven points described below. The methods described in Sections 3.4 and 3.5 are assumed to be known and will be used to some extent.

1. The *camera constant* λL has to be determined by means of the Basic Equation (3.1) $\lambda L = Rd$. For this, a *calibration standard* is used (Section 3.2). A calibration standard is not needed if the diffraction pattern has either low indices or is based on a simple crystal structure, so that it may be possible to use the model concept of the reciprocal lattice (Sections 3.3 and 3.4). In that case, indexing is accomplished by comparison with "basic types" of patterns, as is shown, e.g., in Fig. 3.7 for fcc and bcc lattices. If the pattern to be analyzed is not among these basic types, but the diffracting crystal belongs to the cubic crystal system, the method of R_n ratios (Section 3.5) is used.

If a diffraction photograph contains the superimposed patterns of two different crystallites (e.g., the matrix and a precipitate), the camera constant λL has to be determined as accurately as possible, including its error limits, in order to index the precipitate. However, if the matrix is cubic, the camera constant can easily be determined from the matrix reflections by the methods outlined in Sections 3.3 to 3.5.

2. An assumption can be made that the pattern may originate from a certain crystal structure. In that case, a *complete table is drawn up of those interplanar spacings $d(hkl)$* which are larger than a reasonable minimum value.

3. Each diffraction pattern has *twofold symmetry* (see Fig. 3.8). (The special cases of four- or sixfold symmetry are included in that statement.) Therefore it can be synthesized by vector addition of only two vectors, the *basic vectors*.[8] In Fig. 3.8 they are called $\mathbf{R}_1$ and $\mathbf{R}_2$; they lead from the origin (000) to the *basic reflections*. From the camera constant and the R values of these two basic reflections taken from the photograph

---

[8] In real diffraction patterns these principles sometimes appear to be violated. However, in such cases the patterns really consist of several partial patterns which must be analyzed separately. See details in Chapter 4 (simultaneous presence of different Laue zones).

(including error limits!), the associated experimental *d*-values are calculated by division (Basic Equation).

4. Now one checks in the table of *d*-values whether these two *d*-values (considering their error limits) match any of the theoretical values. Only with cubic structures, where the *d*-values are not closely spaced, is one apt to find an immediate unambiguous coordination. At this stage, only the lattice plane *type* {*hkl*} corresponding to the *d*-value has been found. For complete indexing (*hkl*), i.e., including sequence and sign of the indices, the procedures of subsequent points 5–7 below have to be applied. For complicated noncubic structures with closely spaced *d*-values, these procedures also serve to choose the correct value from several which are indistinguishable, considering the given error limits alone.

5. Tentatively assumed indices for the two basic reflections $(h_1 k_1 l_1)$ and $(h_2 k_2 l_2)$ in Fig. 3.8 are tested with two criteria: the *vector addition rule* and the *angular relations* for all $\mathbf{R}_n$:

a) *Vector addition:* All vectors $\mathbf{R}_n$ leading to diffraction spots can be described according to Eq. (3.4). In an orthogonal coordinate system two vectors are added by adding their components, i.e.,

$$h_3 = h_1 + h_2, \quad k_3 = k_1 + k_2, \quad l_3 = l_1 + l_2.$$

This principle is valid also for oblique-angled coordinates, if these are contravariant and not covariant. In a covariant coordinate system, the vector components are found by projection onto the axes; in a contravariant system the vector components are found by drawing parallels to the oblique-angled axes of the cordinate system.

With this rule the determination of the correct indices (*hkl*) is easily checked. From two spots one finds the third by vector addition and checks, whether the *d*-value for the calculated (*hkl*) derived from the table is in agreement with the actual *R*-value.

Figure 3.9 shows as an example the reflection (34$\bar{3}$) obtained by vector addition in a diffraction pattern of an Al_2Cu precipitate crystal (θ-phase) in an Al-4% Cu alloy. The θ-phase is tetragonal with $a = 0.4864$ and $c = 0.6054$ nm.

b) *Angular relationships:* For the final check of correct indexing, an examination of the *angles* between the lattice plane normals is indispensible. In all crystal systems these angles are identical with the angles between the associated vectors $\mathbf{R}_n$ leading to the point (*hkl*) in the diffraction pattern and in the reciprocal lattice, as discussed in Section 3.3 (see also Fig. 3.4). One checks the tentatively adopted indices with respect to angles and vector addition and changes signs and permutation of the indices until *all angles and all d-values are correct*. Only then the diffraction pattern is internally consistent and correctly indexed.

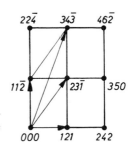

Fig. 3.9. Example for the rule of vector addition for the vectors R_n pointing from the origin (000) to the diffraction spots (hkl).

Equation (3.8) is valid for the angles between lattice planes in the cubic systems. For other crystal systems, the corresponding equations can be found in textbooks on crystallography and x-ray diffraction [15, 16]. A comprehensive and very worthwhile table for cubic crystals is the one by Peavler and Lenusky [9], reprinted in [9a]. It extends to angles between lattice planes of the type {554}.

The indices of *orthogonal reflections* R_n can be quickly checked because the *scalar product* of these vectors must be 0. Specifically for cubic crystals this means that

$$h_1 h_2 + k_1 k_2 + l_1 l_2 = 0. \tag{3.9}$$

For tetragonal or orthorhombic lattices the scalar product is ($h_1 h_2 \mathbf{a}^2 + k_1 k_2 \mathbf{b}^2 + l_1 l_2 \mathbf{c}^2$) with $\mathbf{a}$, $\mathbf{b}$, $\mathbf{c}$ being the lattice parameters.

6. Determination of the crystal normal (zone axis) [uvw]. The direction normal to the plane of the diffraction pattern is designated with [uvw]. It is parallel to the primary beam $\mathbf{P}$. [uvw] coincides with the foil normal only if the foil is exactly normal to $\mathbf{P}$ (compare Section 3.4). The zone axis [uvw] is obtained as the *vector product* of two arbitrary vectors R_n in the plane of the diffraction pattern. The vector components are

$$u = k_1 l_2 - k_2 l_1, \quad v = l_1 h_2 - l_2 h_1, \quad w = h_1 k_2 - h_2 k_1. \tag{3.10}$$

For a quick determination of u, v, w, these components can be represented as subdeterminants of the matrix containing a^*, b^*, c^*, and R_1, R_2.

When designating the two reflections with the subscripts "1" and "2," care has to be taken that in the plane of the diffraction pattern, R_2 is obtained by mathematically positive (counterclockwise) rotation of R_1 by less than 180°. If this orientation relation is observed, [uvw] points upward, i.e., antiparallel to the primary beam $\mathbf{P}$.

Equation (3.10) is valid for all lattice systems, including noncubic ones. However, during the analysis and utilization of indexed diffraction patterns from noncubic lattices, it should be kept in mind that an [hkl] *direction* is not necessarily parallel to the *normal of the (hkl) plane*.

7. Check of completed indexing.

a) *"Rule of increasing or decreasing indices."* The h-values for all diffraction spots located on a straight line must increase or decrease from spot to spot by the same integral amount. (The integers include zero and negative numbers.) The h-values form an arithmetic series. The same rule holds for the k- and l-values (see Fig. 3.9). This simple property of lattices follows from Eq. (3.4).

b) The *scalar product* of the zone axis with *each* individual reflection (hkl) has to be zero, i.e., $uh + vk + wl = 0$.

Finally, it should be pointed out that the strict application of these seven indexing rules constitutes the necessary and sufficient condition for correct indexing.[9] *This is true for all crystal systems.* Especially for noncubic structures the beginner often obtains wrong indices if he does not apply *all* seven rules. The reason is that the d-values of structures with low symmetry are relatively close together and by chance occasionally the d-values, for example, as well as the vector additions for three reflections, appear to be correct. Yet indexing may be wrong, and only a check of the angles will show it. Readers gaining experience by indexing numerous diffraction patterns will acknowledge these facts.

All indexed patterns shown in the other chapters, e.g., Figs. 1.13, 3.7, 3.17, 4.32, etc., are *examples* for the indexing of diffraction patterns.

The indexing of diffraction spot patterns can also be done by computer. For example, Schwab and Griem [10] have described and are using a suitable program for fully automatic indexing following the above seven rules: The input consists of λL, the lattice parameters, and the coordinates of the basic vectors; the indices (hkl) of the diffraction spots and those of the zone axis $[uvw]$ are printed out. In addition, if the first run through did not yield a solution, the program can apply small variations to the lattice parameters, with the aim of finding a solution in this manner. This may be of importance for intermetallic phases, carbides in steels, etc., whose lattice parameters may vary within certain crystallographically imposed limits. Other computer programs for indexing electron diffraction patterns are reported in [22] and [23].

3.7. Correlation of Image and Diffraction Pattern: Magnetic Rotation

After the diffraction pattern is indexed, it has to be *correctly correlated* with the image. Due to the electromagnetic lens fields, the electron beam

---

[9] In very few cases it may happen that two different solutions exist for the same diffraction pattern, but only one solution is correct. This "coincidence ambiguity" is discussed in Section 3.10.

travels along a vertical helix (see derivation in Section 1.3). The "magnetic rotation" thus caused was demonstrated in Fig. 1.4. Since for the image and the diffraction pattern different planes in the EM are imaged, the image and its associated diffraction pattern are *rotated* against each other by a certain angle. This angle φ of the magnetic rotation is a function of the magnification M (Fig. 3.10a). $\varphi = \varphi(M)$ must be determined empirically for each EM. This can be done conveniently with such crystal samples which clearly show in their image certain crystallographic directions or planes (e.g., plate-shaped precipitates on cube faces, or well-defined crystalline cleavage planes or directions for ionic crystals). Typical values for φ are for example 25° for 20,000× and 45° for 40,000× magnification (Elmiskop I).

If φ is known, the image and its diffraction pattern can be correlated as shown in Fig. 3.10b. The diffraction pattern has to be rotated clockwise with respect to the image, and *in addition by 180°*. This latter rotation is due to the fact that (according to Fig. 1.3a,b) the image has undergone a three-stage magnification, while the diffraction pattern is magnified by only two stages; and each magnification stage is accompanied by a 180° rotation. These 180° rotations are *important for the asymmetrical diffraction patterns to be discussed in Chapter 4* (two-beam case with a given diffraction vector **g**). In order not to overlook the additional 180° rotation, it is recommended to always number the negatives in the same corner as, for example, in Fig. 3.10 the numbers 5000 and 5001. The situation described applies to *negatives*. For *positives* (prints or enlargements), a mirror image of Fig. 3.10 has to be used.

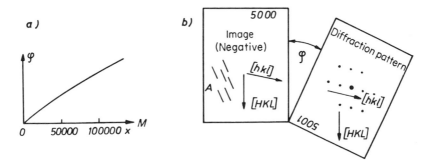

Fig. 3.10. Magnetic rotation ϕ as a function of magnification M. b) Correlation of image and diffraction pattern. The diagram is correct for the photographic negative. (Schematic for Siemens Elmiskop I. For other types of EM, e.g., JEM or Philips EM300, the sign and magnitude of the angle ϕ may be different.)

3.8. Determination of Directions and Planes; Trace Analysis

After an electron diffraction pattern has been indexed, the next task generally is the crystallographic indexing of certain directions and planes in objects seen in the associated TEM image, as, e.g., the directions $[20\bar{2}]$ and $[1\bar{1}1]$ marked in Fig. 4.25. For example, habitus planes of plate-shaped precipitates or stacking fault planes may warrant the indexing of planes; directions would be of interest, e.g., for habitus directions of needle or rod precipitates in Widmannstätten arrangement, or for dislocation lines in parallel alignment. Such an analysis comparing image and diffraction pattern with the aim of determining geometric-crystallographic data of image features is called *trace analysis*. If a plane located obliquely in a thin foil intersects one or both foil surfaces, the line of intersection is called the *trace* of the plane (as in light microscopy).

After the magnetic rotation is taken into account, all indexed lattice plane normals $\mathbf{R}_n(hkl)$ located in the plane of the diffraction pattern can be transferred onto the image by parallel translation. In cubic lattices, this *normal to the plane* $\mathbf{R}_n(hkl)$ is *identical* with *the direction* $[hkl]$ of the same indices. In Fig. 3.10b this transfer of directions lying in the plane of the photographs is shown by the example of the directions $[hkl]$ and $[HKL]$ drawn through rows of diffraction spots.

With noncubic lattices, it has to be kept in mind that the straight lines through the diffraction spots yield only *plane normals* which are *not necessarily* identical with the *directions* of the same indices. However, the directions can be constructed from the totality of the planes. In the following, cubic lattices are meant when "directions" are mentioned. For noncubic lattices, the reader may want to make the necessary changes as an exercise.

Assume that the image shows linear objects in a certain direction which in Fig. 3.10b are marked with A. If these appear to lie parallel to a direction $[h'k'l']$, it cannot be assumed that, in fact, they do lie in that direction. This would be true only if the linear objects were normal to the primary beam (parallel to the foil surface). *One must always take into account the fact that the TEM image is a projection of the foil volume* (albeit often a very thin one).

A foil thickness of 200 nm corresponds to 4 mm at $20,000\times$ magnification. If the image contains comparatively long objects with a true length l, the angle between the objects and foil surface cannot be very large. In any case, it is smaller than $\arcsin(d/l)$. If, however, the linear features are short, the possible effect of projection definitely has to be considered. Although, in that case, an exact analysis of the direction present in the sample volume is possible, it is more complicated since several photographs are necessary. The analysis can be done in the following way.

The linear features in Fig. 3.10b lie on the *image projection* parallel to the direction $[h'k'l']$. The latter is drawn into a *stereographic projection*,

Fig. 3.11a. The edges of the photograph—these are reference directions fixed with respect to the sample—are taken as the main axes of the projection; the center M corresponds to the image normal (primary beam). All directions identified by diffraction spots correspond to points on the circumference of the stereographic projection. This includes $[hkl]$ and $[HKL]$. In both Figs. 3.11a and 3.10b, these two directions are drawn for $hkl = \bar{2}00$ and $HKL = \bar{1}13$, as given in the indexing example of Section 3.5. The only statement that, at this point, can be made about the true direction of the objects A, which in the *image projection* show the direction $[h'k'l']$, is the following: In the stereographic projection, they must lie on the diameter D_1 through the pole $h'k'l'$. This diameter is the possible locus for the direction sought; it is the straight line marked D_1 in Fig. 3.11a. To facilitate the following procedure, this stereographic projection is transformed into a (001) *standard projection*.[10]

In the special example drawn, this is done by first rotating the projection until the $\bar{2}00$ pole in Fig. 3.11a coincides with the $\bar{1}00$ pole of the standard projection in Fig. 3.11b. Then the projection is rolled about the $\bar{1}00$–100 axis until the $\bar{1}13$ pole is located at its appropriate point in the standard projection. The point is marked in Fig. 3.11b.

As a consequence of this transformation, the diameter D_1 in Fig. 3.11a is transformed into a great circle, marked with D_1 in Fig. 3.11b. The direction sought is located on this great circle. To determine it unambiguously, a second pair of photographs (image and diffraction pattern) must now be taken in a different orientation of the sample. This can be done

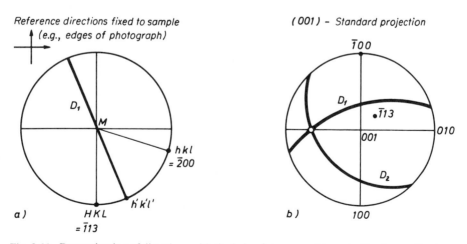

Fig. 3.11. Determination of directions with the help of stereographic projection (generalized case).

[10] The stereographic technique necessary for this transformation is described in more detail in the textbooks on x-ray structure analysis cited in [15]. See in particular [16].

either by selecting a differently oriented grain containing the same linear objects, or by tilting the sample, if a sufficiently large tilt is experimentally possible. From this second set of photographs the independently determined projection of the direction sought is graphically transformed in the same way onto the (001) standard projection. The result is the great circle marked D_2 in Fig. 3.11b. The *intersection* of the two great circles D_1 and D_2 is the unambiguous graphical solution for the unknown direction represented as a pole in the (001) standard projection. This pole is marked with an open circle in Fig. 3.11b. Its numerical indices are obtained by measuring the angular distance of the point from the three {100} poles. (The three indices are in the same ratio as the cosines of the respective angles.)

The method described can take care of *the general case* for the determination of directions. Usually it has to be used only for high-index directions located obliquely in the foil. In practice, the determination of directions often is much simpler: If possible, one photographs image–diffraction pattern pairs of the most important low-index matrix orientations, as e.g., {100}, {110}, and {111}. Using three-dimensional visualization or stereographic projection, directions with not-too-high indices can often be identified without difficulty.

The same goes for the *analysis of planes*. In Fig. 1.13, for example, it can easily be recognized that the plate-shaped precipitates lie on cube faces. The matrix has a [116] orientation. Figure 3.12a also shows platelet precipitates on cube faces; the matrix direction [023] is approximately parallel **P**.

In the general case, an unknown *plane normal* can be indexed in the same way as described for an unknown direction. The orientation of a plane cannot be determined from its trace on an image, because the trace can originate from an infinity of planes, all containing the trace as their zone axis. The straight line at right angles to the trace is the projection of the plane normal. This projection can be analyzed in the same way as was shown for unknown directions; in this way the plane normal is indexed.

Finally, it should be mentioned that in contrast to the *graphical procedures* described here there are also *analytical methods* for the determination of directions and planes (see, e.g., [11]).

3.9. Foil Thickness Determination by the Trace Method

Since, as mentioned before, samples for electron microscopy with 100 kV electrons are transparent only if the sample thickness, depending on atomic weight or mass thickness, does not exceed 50–300 nm, the question

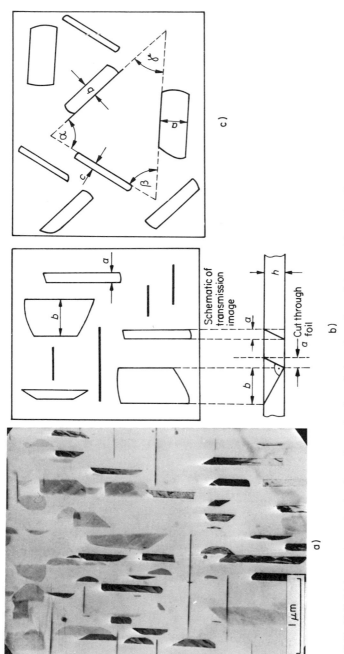

Fig. 3.12. Foil thickness determination from the projection of plate-shaped precipitates. Alloy Al–4% Cu, solution annealed + aged 63 hr at 225°C, θ′ phase. a) Matrix in (hk0) orientation, TEM image. Foil thickness at the upper edge of the photograph 120 nm, at the lower edge 150 nm, 20,000 ×. b) As in a), schematic. c) Case of a general (hkl) orientation, schematic. (After von Heimendahl and Wassermann [12], and von Heimendahl [13].)

arises how to measure such foil thicknesses. A large number of methods is available for this task. Some are based on comparative image brightness I_t of the transmitted beam, i.e., on contrast measurements. These, of course, are applicable only to *amorphous materials* (e.g., carbon or plastic films), since with crystalline materials I_t strongly depends on the orientation of the sample (Bragg diffraction, compare Section 1.6). Many materials are crystalline; and for these, because of their crystalline lattice, specific thickness measurements suggest themselves. (Note that all such thickness measurements, in contrast to macroscopic thickness measurements, are for the *local thickness* of that sample area visible in the EM. These areas, of course, are of main interest anyway.)

Thickness determination is particularly simple in the special case when, e.g., a metal sample contains a second phase in the form of thin platelet precipitates on specific matrix habit planes. The foil thickness can be calculated from the *projection of these precipitates*.

An example is the platelet precipitation of θ' (metastable phase of composition Al_2Cu) in an Al–4%Cu alloy. These precipitates lie on the three {100} planes (the cube planes) of the cubic Al matrix (Fig. 3.12). With proper heat treatment, the approximately circular platelets may have a diameter larger than the unknown foil thickness t. Unless they are parallel or at a very small angle to the foil surface, the platelets usually extend from one surface to the other. Figure 3.12a shows the θ' precipitates [12] in the special case where one of the three matrix cube planes is exactly normal to the foil plane. Thus, an $(hk0)$ matrix plane is oriented normal to the primary beam. The platelets normal to the foil surface appear as thin lines. The line width corresponds to the *platelet thickness* which is here ≤ 10 nm and can be disregarded for purposes of foil thickness determination. The platelets lying on the other two cube planes can be recognized in Fig. 3.12a by their distinctly different widths. However, for each of the cube planes *width* and *contrast* are uniform. The smaller the angle between platelet normal and foil normal, the wider is the projection of the platelets and the lower is their (mass thickness) contrast. Many of the platelets are bordered by two parallel lines. The reason is that before foil thinning the platelet diameter was approximately 10 times as large as the final foil thickness. After thinning, one of the straight lines corresponds to the upper surface of the foil, the other to the lower.

In Fig. 3.12b the three platelet systems are shown schematically with the projected (measured) widths a and b. Underneath is a schematic of a section through the foil and platelets. Since systems lying on cube planes are orthogonal, the thickness t of the foil follows from plane geometry:

$$t = (1/M)(ab)^{1/2},$$

where M is the magnification of the micrograph.

For Fig. 3.12a the calculation yields a foil thickness of 120 nm for the upper edge of the picture, and 150 nm for the lower edge. Thus, such measurements also yield information about the wedge shape of the foil.

When the sample orientation is not $(hk0)$ but an arbitrary (hkl), the three platelet systems appear generally as drawn schematically in Fig. 3.12c. From the projected plate widths a, b, and c, and the angles (see sketch) α, β, γ, the equations for the foil thickness t can be determined with the help of a three-dimensional model [13]:

$$t = \frac{1}{M}\left(\frac{ab}{\cos \gamma}\right)^{1/2}, \text{ or } t = \frac{1}{M}\left(\frac{ac}{\cos \beta}\right)^{1/2}, \text{ or } t = \frac{1}{M}\left(\frac{bc}{\cos \alpha}\right)^{1/2}.$$

These equations show that t can always be determined from measurements on only two platelet systems. If all three systems are used, three independent values are obtained, thus permitting a determination of the accuracy of the method.

For needle- or rod-shaped orthogonal precipitates, equations have also been developed [13]. Finally, by applying solid geometry, foil thickness equations can also be derived for cases of platelets or needles lying on arbitrary but known habit planes.

Although this method is quite exact, in the form described it depends on the existence of suitable precipitates. However, it can easily be modified to include single-phase alloys or pure metals if *trace analysis* as described in the last section is used. A sample area under investigation in the EM can be locally heated with the electron beam (see Chapter 1). This often produces a slight plastic deformation. The heat stress causes some of the dislocations on {111} planes to slip through the foil. During passage of the dislocation, the contamination film (Section 1.8) on both foil surfaces is cracked. Consequently two parallel lines, also called *slip traces*, remain on the surfaces. They are the traces of those {111} planes which were "active." An example is seen in Fig. 3.13a at G. The dislocation itself is visible as a bowed line connecting the two ends of the traces only if it stops within the field of view. In the case of Fig. 3.13, this has occurred at the grain boundary K which has stopped the dislocation.

The associated diffraction pattern, Fig. 3.13c, shows that the foil is parallel (110) (see Fig. 3.7a, upper left). The directions taken from the diffraction pattern have been transferred to the standard stereographic projection, Fig. 3.13d. From this projection it can be seen that the trace G must be a (111) or (11$\bar{1}$) trace, and that its normal N forms an angle of $\alpha = 35.3°$ with the [110] picture plane normal **P**. Fig. 3.12e shows schematically a section through the foil with the positions of G, N, P, and α. The foil thickness t follows from tan $\alpha = t$/trace width. At the place G the

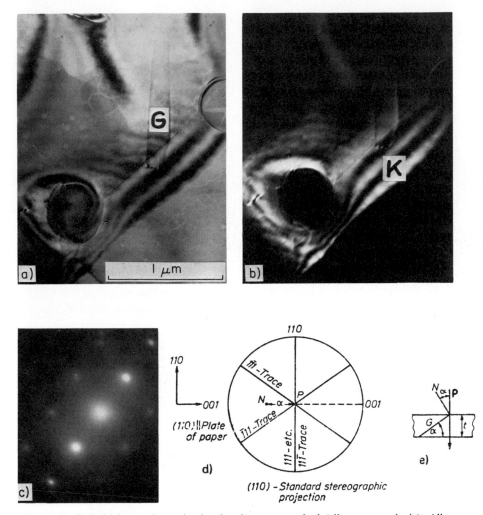

Fig. 3.13. Foil thickness determination by the trace method (slip trace analysis). Alloy Al–4% Cu: a) bright field micrograph, 30,000×, b) dark field micrograph with the upper right (111) reflection seen in c), c) diffraction pattern, d) associated stereographic projection, e) section through foil. (Photographs by Sieberer.)

trace width is 170 nm. With $\alpha = 35°$, the foil thickness t is calculated to be $t = 0.70 \times 170 = 120$ nm.

Figure 3.13a shows that the foil is wedge-shaped (as was also true for Fig. 3.12a): The trace G becomes wider toward the bottom, i.e., the foil there becomes thicker.

For the analysis described, either a bright field or a dark field micrograph can be used. The juxtaposition of both in Fig. 3.13 shows an advantage of the dark field image: Since imaging occurs only with the radiation reflected from specific crystallographic planes, dirt (mass contrast!) is not imaged, while such dirt is visible in the bright field image next to the slip trace G.

If the foil contains no features like dislocations or precipitates on habit planes, the following variant methods for foil thickness (t) determination are available. If there are any objects extending from one surface of the specimen to the other, t can be obtained by *tilting* the specimen from the position exactly normal to the primary beam (zero position) to a large, known angle α around a tilt axis normal to the primary beam. Since the tilting produces significant and measurable changes in the (projected) images of the "objects," these changes can be used to determine t (Vingsbo [24]).

Von Heimendahl [25, 26] has extended this method to the most general case, no longer dependent on any features within the specimen or at the surfaces. *Latex balls* are applied to both specimen surfaces, such as is done, e.g., for magnification calibration or for marking certain areas. The

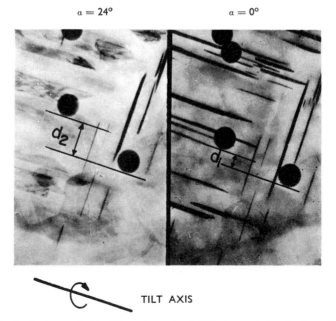

Fig. 3.14. Latex balls applied to foil surfaces for foil thickness determination. Electrolytically thinned Al–4% Cu alloy, foil thickness 324 nm ± 4%, 25,600×.

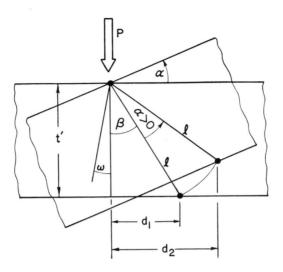

Fig. 3.15. Geometric relations for foil thickness determination with latex balls. Cross section through hypothetical foil of thickness t' ($=$ real foil thickness t + diameter D of latex balls). Tilt axis normal to plane of paper. α = tilt angle; l = projection of ball distance onto plane of paper; d_1 and d_2 are projections of l onto the plane of the paper before and after tilting, respectively; P = primary beam. (For β and ω see von Heimendahl [25].)

Latex milk containing the Latex balls is commercially available, the balls having an exact and uniform size (one species: D = 234 ± 2.6 nm). Figure 3.14 shows a thin foil before and after tilting α = 24°. The two balls marked by the distance d are on different surfaces. This can be checked by tilting from zero first to the one, then to the other direction: if the balls are on the same side, the projected widths d_1 and d_2 would both *decrease* during tilting. If, however, they are on different sides (this is what is needed for the method), one of the two projected widths must *increase*, see Fig. 3.15. This figure refers to a hypothetical foil cross section of thickness t + $(2D/2)$, since the method refers to the centers of the Latex balls being measured. The equation for the foil thickness t

$$t = t' - D = (d_2 - d_1 \cos \alpha)/\sin \alpha - D$$

is easily derived [25]. It refers to the situation of Fig. 3.15, i.e., $d_2 > d_1$, after tilting, with $\alpha > 0$. If the foil is tilted into the opposite direction, an analogous equation is given in von Heimendahl [25]; also the accuracy of the method is discussed there, the latter depending on the angles α and ω. For details, see the original paper.

3.10. Uniqueness (Unambiguity) of Orientation Determination

3.10.1. The 180° Ambiguity

Because of Theorem 1, the reflecting lattice planes are all nearly parallel to the primary beam P. When the crystal is rotated by 180° around the axis P,[11] all these lattice planes will be placed in positions identical to the positions before rotation. Therefore both diffraction patterns, created before and after the 180° rotation around P, are identical. In general, however, the associated crystal orientations are completely different, Fig. 316A. *This 180° ambiguity is a fundamental property of all crystal systems.* It corresponds to a mutiplication with -1 of all diffraction indices (hkl), while the indices of the foil normal $[uvw]$, i.e., the crystal axis parallel to P, remain unchanged.

One hundred eighty degree ambiguity is present only if the diffraction pattern contains the reflections of only *one* Laue zone (see Chapter 4). In that case, the spot pattern of the diffraction pattern has twofold symmetry (centrosymmetrical with respect to the origin), i.e., after 180° rotation it remains the same (Fig. 3.16A).

For the special case of cubic lattices, the 180° ambiguity is illustrated by representation in the stereographic projection, Fig. 3.16B. The reference directions of the projection are fixed in the sample, e.g., the primary beam P and two orthogonal directions A and B normal to P. The two basic reflections $h_1k_1l_1$ and $h_2k_2l_2$ of the diffraction pattern are located on the circumference of the projection, i.e., normal to P. If the reflections are indexed as in case I, the three cube poles could, for example, have the positions indicated in Fig. 3.16B. (The cube poles can be determined from the angular distances of the two basic reflections to the cube poles (100), (010), and (001).) The position in space of the unit cube can easily be visualized from the location of these cube poles. Because of the twofold symmetry of the diffraction pattern one could also have indexed the two reflections as in case II, i.e., one could have multiplied the indices of case I with -1. This corresponds to a rotation of the crystal by 180° around P.

The 180° ambiguity should not be confused with the "arbitrary starting option" when choosing the indices for the first-indexed reflection from lattices of high symmetry, especially cubic lattices. When naming the first-indexed reflection of general orientation $(h_1k_1l_1)$ in cubic lattices,

---

[11] Strictly speaking: around the zone axis $[uvw]$ normal to those reciprocal lattice points which generate the diffraction pattern. For the problem of ambiguity, however, the difference between $[uvw]$ and P is immaterial.

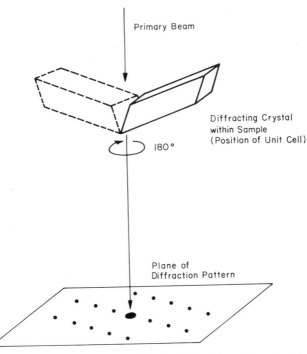

Fig. 3.16A. 180° ambiguity of spot patterns. The dashed position of the unit cell of the crystal is generated from the solid-line position by 180° rotation around the primary beam. (Drawing for the generalized case of a triclinic lattice.) Both positions produce the same diffraction pattern (bottom). (From von Heimendahl [17].)

one can choose any of the 48 possibilities which are obtained by multiplying the six permutations of the *hkl* with the eight sign combinations $+++$, $-++$, $+-+$, $++-$, $---$, $+--$, $-+-$, $--+$. These 48 indexing possibilities do not, of course, correspond to different physical realities, in contrast to the ambiguity, but they are distributed, 24 each, to case I and case II of Fig. 3.16B: each cubic coordinate system in the two different cases can be indexed arbitrarily in 24 different ways, depending on which axis one wishes to call (100), ($\bar{1}$00), (010), etc. (It is customary to use only right-hand systems.) Choosing the indices of the first reflection ($h_1 k_1 l_1$) also fixes the indices of the three main axes. After the first reflection is named, indexing of all other reflections, of course, has to be "consistent," i.e., follow the indexing rules laid down in Section 3.6, to avoid mistakes.

   In contrast to the arbitrary starting option, the 180° ambiguity occurs in all crystal systems. There are several ways to resolve this ambiguity:

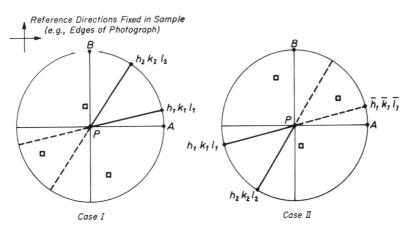

Reference Directions Fixed in Sample
(e.g., Edges of Photograph)

Case I

Case II

Fig. 3.16B. 180° ambiguity for the special case of cubic lattices; stereographic projection (reference system fixed in sample). □ poles of cube faces. (From von Heimendahl [17].)

As an exception, a diffraction pattern is *unambiguous,* if the primary beam P is parallel to a crystal axis of *n-fold symmetry with n being even.* In such a case, rotation by 180° around P produces the identical orientation. In the cubic lattices this occurs only for the following orientations:

$P\|\langle100\rangle$ (fourfold symmetry)   and $P\|\langle110\rangle$ (twofold symmetry).

All other axes are of a symmetry with n being odd, e.g., $\langle111\rangle$ has threefold symmetry $\langle112\rangle$ and all other axes have only onefold symmetry.

Additional methods for resolving the 180° ambiguity will be discussed in Section 4.12b, after some of the prerequisites have been introduced.

3.10.2. Coincidence Ambiguity

The two crystal orientations constituting the 180° ambiguity are symmetrical to P and occur in all crystal systems. A different type of ambiguity occurs only in highly symmetrical lattices, cubic in particular: in certain, relatively rare cases, the same diffraction pattern can be indexed in two completely different ways, corresponding to two crystal orientations which are completely different and are not characterized by any symmetry relation. These two sets of indices formally agree with the indexing rules. Yet only one is correct, i.e., corresponds to the real orientation of the crystal.

An example is given in Fig. 3.16C; it is the diffraction pattern of an aluminum foil taken with 100 kV electrons. A check of the rules in Section

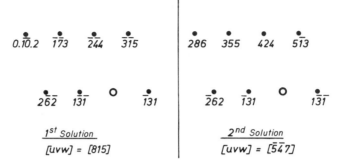

Fig. 3.16.C. Case of a coincidence ambiguity. The two spot patterns are identical, yet they can be produced by completely different crystal orientations of a fcc lattice. (From von Heimendahl [17].)

3.6 shows that both sets of indices are internally consistent and therefore formally correct. A necessary prerequisite for this kind of ambiguity is the existence of at least *one* reflection which can be indexed in two completely different ways, (hkl) and $(h'k'l')$, respectively, but for which the sum of the squares of the indices $h^2 + k^2 + l^2$ and $h'^2 + k'^2 + l'^2$ is the same, i.e., they *coincide,* as for $\bar{1}73$ and 355, and for $0.\overline{10}.2$ and 286 in Fig. 3.16C. Therefore this kind of ambiguity is called "coincidence ambiguity." All other reflections in the two indexing possibilities belong to the same type $\{hkl\}$, i.e., they are distinguished only by permutation and sign.

Unfortunately, there isn't enough space for a more detailed description of the conditions occurring with coincidence ambiguity. The reader should consult the original papers [17–19]. Samudra *et al.* [19] contains *a complete listing of all cases of coincidence ambiguity* for cubic and tetragonal lattices up to the values for $u^2 + v^2 + w^2 = 350$ and $h^2 + k^2 + l^2 = 200$. In other crystal systems, the coincidence ambiguity does *not* occur. A method for resolving this ambiguity will be described in Section 4.12.

3.11. Kikuchi Lines

So far the *accuracy* of orientation determination by the analysis of a single-crystal spot pattern has not been discussed. The next section will show that, unfortunately, that accuracy in general is not very great unless additional special procedures are applied. The accuracy is in the order of magnitude of $\pm 5°$; in individual cases it can be better or worse. Experimentally this can be verified by tilting the sample in the EM. A few degrees of tilt will not change the geometrical relations of the diffraction spots.

Special methods, however, are available for determining the orientation with an accuracy of ± 1° or better. One of these makes use of the so-called *Kikuchi lines* [14]. When one investigates diffraction phenomena as a function of foil thickness, by examining areas away from the thin sample edge, it will be observed that the *diffraction spots* become progressively weaker and that instead a system of parallel, bright and dark lines appears. Figure 3.17 is an example for these *Kikuchi lines*.

Origin of Kikuchi lines

The scattering of electrons (see Section 1.6) in thicker foils is not only *elastic* but, increasingly with increasing thickness, also *inelastic*. While elastic scattering in crystalline samples produces the *separate* maxima of Bragg diffraction, inelastic scattering occurs *diffusely* in all directions with maximum intensity in the direction of the primary beam and intensities decreasing with increasing angle. This is illustrated in Fig. 3.18. Within the sample, the primary beam I_0 produces everywhere along its path a cone of inelastically scattered rays. For the following argument, the assumption is made that the *wavelength* is increased only insignificantly as a consequence of the energy loss during inelastic scattering (i.e., the change can be ignored). Under this assumption, the inelastically scattered electrons can act as new *"primary beams."* (They are set in quotation marks to distinguish them from the real primary beam). These scattered beams on their part can cause *Bragg reflection* at lattice planes.

To understand the effect of this type of Bragg reflection, three crucial facts have to be considered.

a) First, each lattice plane, such as the one marked (*hkl*) in Fig. 3.19, produces a Bragg interference *line,* not a spot. This is due to the fact that, as mentioned, *cones of rays* are produced along the path of the primary beam I_0 in the sample, and these rays are "primary beams" which irradiate the lattice planes not only from one, but *from all directions.* Out of all these "primary beams," the lattice plane reflects, because of the Bragg condition, only those rays which enter from and exit with the angle θ, but at the same time lie symmetrical to the lattice plane, i.e., on a conical surface. Because θ is so small, the intersection of the photographic plate with the conic section of this interference appears to be a straight line on the plate. The line is normal to the plane of the paper in Fig. 3.19.

b) Second, the *scattered intensity* of the inelastically scattered electrons constituting the "primary beams" decreases with increasing angle,

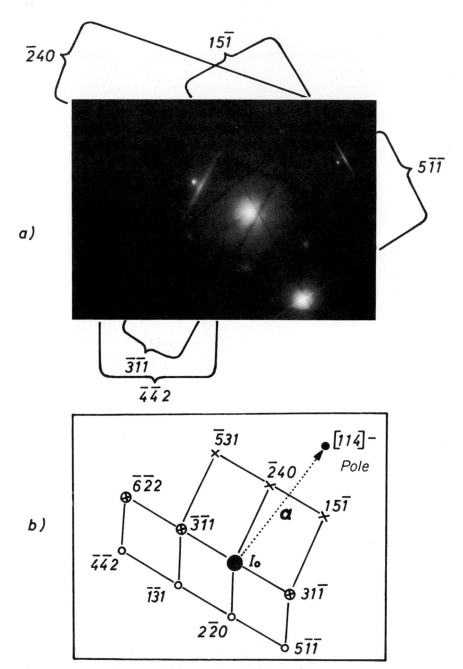

Fig. 3.17ab. Kikuchi lines. a) Five different pairs of Kikuchi lines in a diffraction pattern of aluminum foil, 100 kV. b) Indices of the spot pattern. Reflections of two Laue zones are present: 1st zone with zone axis [114], open circles; 2nd zone with zone axis [217], x. Indices are unique in the sense of Section 4.12.

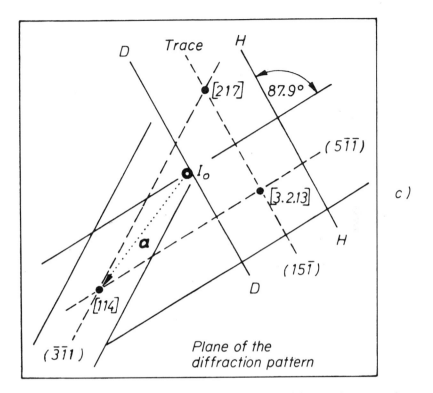

Fig. 3.17c. Construction of traces of a few lattice planes (dashed lines) and zone axes from Fig. 3.17a. *H* are light, *D* are dark Kikuchi lines.

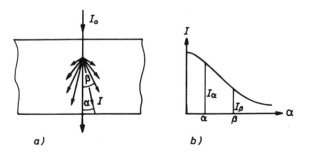

Fig. 3.18. Inelastic scattering in the sample. a) Cross section through the sample, b) scattered intensity of inelastically scattered radiation, schematically.

as shown in Fig. 3.18b. Therefore:

$$I_\alpha > I_\beta \qquad \text{if } \beta > \alpha. \tag{3.11}$$

c) Third, each nearly vertically positioned lattice plane in the foil (only these reflect according to Theorem 1, Section 3.1) is irradiated *from both sides* by the "primary beams." (The main primary beam I_0 has a finite width when compared with the dimensions around (*hkl*).)

These three factors together explain the generation of the Kikuchi lines. The lattice plane (*hkl*) in Fig. 3.19 receives intensity from the "primary beam" I_α at the Bragg angle θ, where α is the angle (I_α, I_0). According to a), an interference line would be produced on the image screen at location H. The intensity of the radiation diffracted from I_α toward H is fI_α, where f is the constant ratio of diffracted and primary intensity ($f < 1$). From its other side, the same lattice plane (*hkl*) receives at the angle θ intensity from the "primary beam" I_β, with $\beta = \text{angle}(I_\beta, I_0)$. According to a) this diffraction produces an interference line at the location D on the image screen. The intensity diffracted from I_β toward D is fI_β. How-

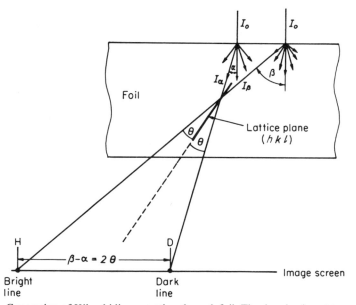

Fig. 3.19. Generation of Kikuchi lines; section through foil. The drawing is not to scale; the angles α, β, θ in reality are very small, and the distance to the image screen is very large compared to the foil thickness.

ever, fI_β is not the total intensity at D. One has to take into account that the diffusely scattered intensity (Fig. 3.18b) illuminating the screen at D has lost *that* intensity fI_α which has been reflected by the lattice plane into the new direction H (changed by 2θ from its original direction). Therefore, fI_α has to be *subtracted* at D. Analogous arguments show that at H intensity fI_β has to be subtracted. The end result of Bragg reflection at the lattice plane is a *superposition* of the following intensities, on top of the continuously decreasing scattered intensity from Fig. 3.18b:

in direction H: $fI_\alpha - fI_\beta > 0$ because of Eq. (3.11)
in direction D: $fI_\beta - fI_\alpha < 0$ because of Eq. (3.11).

Therefore, at H a *bright line* of intensity $f(I_\alpha - I_\beta)$ is produced, while at D a *dark line* appears with an intensity *decreased,* with respect to the background, by the amount $f(I_\beta - I_\alpha)$. The terms "bright" and "dark" refer to the image screen or the positive print; contrast is reversed for the negative (film or plate).

Figure 3.19 also shows that the pair of bright and dark lines, H and D, which are normal to the plane of the paper, are separated exactly by the amount $2\theta = \beta - \alpha$. The extension of the lattice plane (hkl) towards the image screen (the dashed line), i.e., the *trace* of the plane, is the *median or symmetry line* of the two lines H and D. The symmetry lines are also drawn in Fig. 3.17c. The latter shows the pairs H and D and the associated traces (dashed lines) of three lattice planes taken from the example in Fig. 3.17a.

Indexing of Kikuchi Lines, Bragg Case, and Laue Case

Kikuchi lines are *indexed* according to the same rules applicable to spot patterns, except that the distance 2θ between the two lines of a pair replaces the vector $\mathbf{R}_n$. In actual diffraction patterns, *diffraction spots* often are present in addition to Kikuchi lines. The spots are indexed first, as, e.g., in Fig. 3.17b. The stronger Kikuchi lines can be coordinated with their respective diffraction spots, i.e., the spots with the same indices, as follows: The distance 2θ between the lines of a pair on the one hand, and the distance diffraction spot–origin ($\mathbf{R}_n$) on the other, have to be identical. The bright line passes near or directly through the diffraction spot and is *normal to* $\mathbf{R}_n$. Its brightness is the greater, the closer it is to the diffraction spot (compare the reflections and lines $\overline{3}11$, $15\overline{1}$, and $5\overline{1}\overline{1}$ in Fig. 3.17a). If the bright line passes through the center of the spot, as is the case for $5\overline{1}\overline{1}$, $\alpha = 0$ (Fig. 3.19). This can be seen by comparison with Fig. 3.2. *In this case, the Bragg condition for the respective lattice plane (hkl) is exactly met also for the primary beam I_0. Therefore it is called the exact Bragg case, or short "Bragg case" for the lattice plane in question.* The corresponding diffraction spot *hkl* has a particularly high (maximum) intensity. This is true for the $5\overline{1}\overline{1}$ reflection in Fig. 3.17. If the intensity of all other reflection spots is negligible compared with the first, the condition is called a *two-beam condition*. This means that aside from the primary beam only this one other diffracted beam is of importance (example Fig. 4.13).

If, however, all reflecting lattice planes are exactly parallel to the primary beam I_0, this is called the symmetrical or Laue case. In this case all diffraction spots have nearly equal intensity, as, e.g., in Fig. 3.5b and, schematically, Fig. 4.11. The Kikuchi lines are also symmetrical to I_0. The Laue case is a special case among those in which several diffracted rays of comparable intensity are present: *multibeam condition.*[12]

Determination of Exact Crystal Orientation with Kikuchi Lines

With Kikuchi lines the orientation of the imaged crystal with respect to the primary beam I_0 can be determined very accurately (e.g. von Heimendahl [14]). The *intersection* of any two lattice plane traces, drawn in Fig. 3.17c, (indices in parenthesis) is their common *zone axis* (indices in square brackets). The indices of the zone axis are obtained as the vector product of the indices of the two planes, according to Eq. (3.10). From the diffraction pattern the exact position of the primary beam I_0 with respect to the indexed zone axes is obtained as follows:

If the photograph contains Kikuchi lines of more than one zone, as is the case in Fig. 3.17c, the determination is especially simple and, moreover, unambiguous (see Section 4.12). Three known zone axes are present as a triangle, as in Fig. 3.17c. They are transferred to a (001) standard stereographic projection where they form an analogous triangle. The relative position of I_0 with respect to the zone axes is also transferred from Fig. 3.17c to the stereographic projection. With these steps the orientation has been determined graphically. Numerically it is obtained from the angles between I_0 and the three {100} poles. The ratios of the cosines of the angles is the same as those of $h:k:l$. In the example, the exact orientation of I_0 is [25, 17, 95].

If Kikuchi lines of only *one zone* are present, i.e., all lines have only *one* intersection, the exact orientation can also be determined. Assume that only the pole [114] in Fig. 3.17c is present. The deviation **a**, i.e.,

---

[12] See discussion of Laue case and Bragg case in Fig. 4.8b.

length and direction, of this known pole from the primary beam I_0 can be measured. The tilt ϵ away from the [114] position follows from Eq. (3.12)

$$\epsilon = |\mathbf{a}|/L, \tag{3.12}$$

where L is the camera length to be determined from the diffraction pattern. This equation follows directly from Figs. 3.2 and 3.19. In the example, $a = 38$ mm, $L = 545$ mm, $\epsilon = 0.070$ or $4.0°$. To determine the direction of $\mathbf{a}$, one has to keep in mind that the zone axes (poles) drawn in Fig. 3.17c point *downward*, from the sample to the photographic plate, as seen in Fig. 3.19. In the usual analyses, however, directions are positive in the *upward* direction (see point 6 of the rules in Section 3.6). Figure 3.20 is a perspective drawing to clarify these facts. The tilt axis *TA* is found as the direction normal to $\mathbf{a}$ in the diffraction pattern Fig. 3.17b. This information is all one needs to solve the problem: In the stereographic projection with reference directions A, B, and P fixed in the sample (as in Fig. 3.16B), the exact positions of the poles *TA* and [uvw] are plotted. From two known poles all other poles (directions) can be found. In particular, the three $\langle 100 \rangle$ directions of a cubic lattice can be plotted in relation to A, B, and P, and, if desired, the stereographic projection with reference directions A, B, and P can be transformed into a $\langle 100 \rangle$ standard projection [20]. In addition to this graphical solution, an analytical (numerical) solution exists for determining the orientation from a single Kikuchi pole. See the original paper by Pumphrey and Bowkett [21].

If spots or lines of only one zone are present, an exact orientation determination of I_0 can be carried out as described. However, there remains the ambiguity of the solution as discussed in Sections 3.10 and 4.12.

For other analytical methods to exactly determine orientations from Kikuchi lines, see Schwartzkopff [11].

Calibration of Tilting Stages

If a stereo or tilting stage is used to *tilt* a sample around an axis normal to the primary beam, *the system of Kikuchi lines shifts* parallel to itself. The *position of the diffraction spots*, however, remains unchanged, although their intensity may change (see Chapter 4). If Fig. 3.19 has been understood, the line shift should be clear. The system of Kikuchi lines is rigidly connected with the crystal lattice. The shift vector $\mathbf{a}$ is proportional to the tilt angle. These facts can be used to *calibrate stereo* and *tilting stages* [14]. The tilt angle can be determined experimentally as the

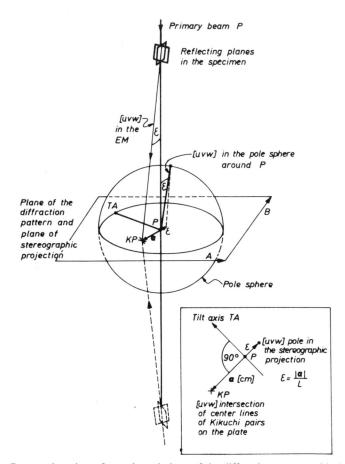

Fig. 3.20. Perspective view of sample and plane of the diffraction pattern with the Kikuchi pole *KP*. Generalized case. Insert: Location of *KP*, *P*, *TA*, **a**, and ε in the plane of diffraction pattern and stereographic projection, respectively. (The magnetic rotation of the electron beam, Section 3.7, is neglected in this figure. It should be taken into account if the orientations of the real reflecting planes in the sample are to be determined.) (From von Heimendahl [20].)

ratio $|\mathbf{a}|:L$, and for calibration purposes this value can be compared with the manufacturer's scale on the stage.

As a numerical example, for the (Siemens) Elmiskop I a Kikuchi line shift of $|\mathbf{a}| = 1$ cm corresponds to a tilt angle of approximately 1° [14].

It should also be mentioned that the *sharpness* (width) of individual Kikuchi lines is a sensitive measure of the crystal perfection. It follows from Fig. 3.19 that the lines are the wider (indistinct), the larger the deviation of the lattice plane (*hkl*) from an ideal plane. For this reason, Kikuchi lines are used to judge the quality of crystals. However, more frequently, *Kossel lines* are used, which are the x-ray analogs of Kikuchi lines.

References

1. J. B. Cohen, "Diffraction Methods in Materials Science," Macmillan Series in Materials Science. Macmillan, New York, 1966.
1a. G. Thomas, "Transmission Electron Microscopy of Metals". Wiley, New York, 1966.
2. ASTM Card File, now called Joint Committee on Powder Diffraction Standards Card File, 1916 Race St., Philadelphia, Pennsylvania (Reg. U.S. Pat. Off.).
3. L. Reimer, "Elektronenmikroskopische Untersuchungs- und Präparationsmethoden," 2nd. ed., Springer, Berlin and New York, 1967.
4. K. Schwartzkopff, *Z. Metall.* **58**, 243 (1967).
5. P. B. Hirsch, A. Howie, R. B. Nicholson, D. W. Pashley, and M. J. Whelan, "Electron Microscopy of Thin Crystals." Butterworths, London, 1971.
6. E. Eichen, C. Laird, and W. R. Bitler: Reciprocal Lattice Diffraction Patterns for f.c.c., b.c.c., and h.c.p., and Diamond Cubic Crystal Systems. Special Rep. by the Scientific Laboratory, Ford Motor Co. Dearborn, Michigan.
7. R. L. Nolder and G. Thomas, USAEC Rep. UCRL 10227. Univ. of California, Berkeley, California.
8. W. R. Roser and G. Thomas, *J. Sci. Instrum.* **35**, 613 (1964).
9. R. J. Peavler and J. L. Lenusky, Angles Between Planes in Cubic Crystals, IMD Special Rep. Ser., No. 8, available from The Metallurgical Society—AIME—29 West 39th Street, New York, New York. The table is reprinted in [9a].
9a. K. W. Andrews, D. J. Dyson, and S. R. Keown, "Interpretation of Electron Diffraction Patterns," pp. 124 - 147. 2nd ed., Adam Hilger, Ltd., London, 1971.
10. W. Griem and P. Schwaab, *Pract. Metallogr.* **14**, 389 (1977).
11. K. Schwartzkopff, *Metall* **23**, 119 (1969).
12. M. von Heimendahl, and G. Wassermann, *Z. Metall.* **53**, 275 (1962).
13. M. von Heimendahl, *J. Appl. Phys.* **35**, 457 (1964).
14. M. von Heimendahl, W. Bell, and G. Thomas, *J. Appl. Phys.* **35**, 3614 (1964).
15. *Some books and tables for X-ray diffraction analysis:*
    a) B. D. Cullity, "Elements of X-Ray Diffraction," 2nd ed. Addison-Wesley, Reading, Massachusetts, 1978.
    b) R. Glocker, "Materialprüfung mit Röntgenstrahlen," 5th ed., Springer, Berlin and New York, 1971.
    c) W. B. Pearson, "A Handbook of Lattice Spacings and Structures of Metals and Alloys," Vol. I. Pergamon, Oxford, 1964; Vol. II, 1967.
    d) K. Sagel, Tabellen zur Röntgenstrukturanalyse, Vol. 8, *in* "Anleitung für die chem. Lab. Praxis," Springer, Berlin and New York, 1958.
16. O. Johari and G. Thomas, "The Stereographic Projection and its Applications," Wiley (Interscience), New York, 1969.
17. M. von Heimendahl, *Pract. Metallogr.* **8**, 279 (1971).
18. B. J. Duggan and R. L. Segall, *Acta Metall.* **19**, 317 (1971).
19. A. V. Samudra, O. Johari, and M. v. Heimendahl, *Pract. Metallogr.* **9**, 516 (1972).
20. M. von Heimendahl, *Phys. Status Solidi* (a) **5**, 137 (1971).
21. P. H. Pumphrey and K. M. Bowkett, *Phys. Status Solidi (a)* **2**, 339 (1970).
22. G. H. Olsen and W. A. Jesser, *Mater. Sci. Eng.* **5**, 135 (1970).
23. R. A. Ploc and G. H. Keech, *in* "Microscopie Électronique," Vol. II, p. 203 (*Conf. Grenoble*). Société Francaise de Microscopie Electronique, 1970.
24. O. Vingsbo, in "Microscopie Électronique," Vol. I, p. 325 (*Int. Conf. Grenoble*). Société Francaise de Microscopie Electronique, 1970.
25. M. von Heimendahl, *Micron* **4**, 111 (1973).
26. M. von Heimendahl, V. Willig, and J. A. Heywood, *Philips Bulletin* EM 112-(1979)/1, Eindhoven, Netherlands, 1979.

4. Contrast Theory and Applications

In Chapter 1, the effects on image formation of the fundamental processes: *scattering* and *diffraction* of the electrons in the sample, were examined qualitatively. The most important result was found to be the following: Image contrast of a given sample area is due to the intensity I_r lost by scattering and diffraction from the primary beam intensity I_0. I_r is intercepted by the objective aperture and therefore subtracted from I_0. The larger the diffracted or scattered intensity I_r is, the smaller the transmitted intensity $I_d = I_0 - I_r$ is, i.e., the darker the image at that point. Diffraction intensity and image brightness are mutually complementary; therefore quantitative calculation of the first also determines the value of the second.

During electron irradiation of crystalline samples, diffraction patterns are produced. In Chapter 3, the *geometry* of these patterns, i.e., the *position* of the diffraction spots, was discussed in all its details. In Chapter 4, the relative *intensity* I_r of the diffraction maxima will be calculated. Thus it will be possible to interpret the contrast effects from crystalline samples (only crystals are the subject of this chapter). In particular, crystals with different kinds of *lattice defects* (dislocations, grain boundaries, precipitates, stacking faults, twins, etc.) will be treated. At the same time, the calculation of the intensities I_r will provide an understanding for a number of questions left open in Chapter 3, concerning details of the geometry of observed diffraction patterns.

4.1. Kinematical and Dynamical Theory; Concepts

For the mathematical treatment of an electron wave passing through a crystal, two approximations, not equally good, are known. They are called the *kinematical* and the *dynamical theory*.

Kinematical theory

As mentioned in Chapter 1, according to the Huygens–Fresnel principle all atoms can be thought of as sources of *secondary waves* moving into space in all directions. These secondary waves have the same wavelength as the primary wave (direction k_0 in Fig. 4.1). For any direction k, chosen as an independently variable measuring direction, the intensity scattered from all atoms into this direction obeys the law of superposition of wave trains of equal *wavelength* λ, but of different *phases* and eventually of different *amplitudes*. The different phases arise from the *path differences* which are caused by the change of wave direction at the atoms. In Fig. 4.1 only two such atoms are shown. The *special case* of maximum intensity, i.e., superposition of waves with phase differences which are integral multiples of λ, was treated in Chapter 1 as the derivation of the Bragg equation.

With the kinematical theory, the *general case* can be treated, i.e., the calculation of the intensity I_r scattered from all atoms of a crystal K (Fig. 4.2) into the arbitrary direction k. This intensity is proportional to the square of the amplitude (A^2) that results from the superposition of the secondary waves emanating from the atoms. From wave physics it is known that the resultant amplitude A of several waves can be written as a complex expression as follows:

$$A = \sum_n f_n e^{i\varphi_n}, \tag{4.1}$$

where φ_n is the phase and f_n the amplitude of the nth wave. In Section 4.4 a graphic interpretation of Eq. (4.1) will be shown. This is the so-called amplitude–phase diagram which is useful for many practical applications.

The amplitude f_n of a wave scattered from a *single atom* into the direction k is proportional to the amplitude A_0 of the incident wave and inversely proportional to the distance R between scattering center and point of observation:

$$f_n = (A_0/R) f(\theta, Z).$$

In this Eq. $f(\theta, Z)$ is called the *atomic scattering amplitude;* it has the dimension [nm]. $f(\theta, Z)$ is a function of the scattering angle θ $(=$ Bragg

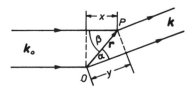

Fig. 4.1. Scattering of an electron wave at two atoms O and P.

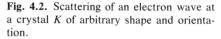

Fig. 4.2. Scattering of an electron wave at a crystal K of arbitrary shape and orientation.

angle = half of the angle between $\mathbf{k}_0$ and $\mathbf{k}$ in Fig. 4.1) and of the atomic number Z; $f(\theta,Z)$ increases with decreasing scattering angle and with increasing atomic number. Qualitatively these relations are clear on the basis of the mechanism of elastic scattering (Fig. 1.6). Quantitatively they can be calculated, e.g., according to the statistical model of Thomas–Fermi–Dirac. The necessary tables for $f(\theta,Z)$ can be found in [1,2]. As an example: for aluminum $f(\theta,Z)$ is 0.61 nm for $\sin \theta/\lambda = 0$, and drops to 0.025 nm for $\sin \theta/\lambda = 1.00$. It should be pointed out that $f(\theta,Z)$ for x-rays is qualitatively analogous to the corresponding $f(\theta,Z)$ for electrons, but not quantitatively the same. The reason is the different nature of the scattering process, see Table 3.1. For x-rays, the atomic scattering factor $f(\theta,Z)$ is dimensionless, since it is defined as the ratio of scattering amplitude of the atom over scattering amplitude of a single shell electron.

The different phases φ_n are due to the different locations of the individual atoms in the crystal. The exact calculation of the resultant amplitude A, with the help of Eq. (4.1), will be given in Section 4.2.

Dynamical theory

For the important and, in practice, frequent case where the sample is more or less exactly in Bragg reflection position, the summation in Eq. (4.1) corresponds to a summation of the waves diffracted ("reflected")[1] at the individual lattice planes, resulting in a single reflected wave. The square of its amplitude is proportional to the reflected intensity I_r. However, in this description two effects have not been taken into account[2]:

a) While the primary beam I_0 penetrates into the crystal, it is continuously weakened because of Bragg reflection. Thus, the lattice planes in deeper locations add a smaller amount to the reflected intensity than do the upper planes ("extinction").

b) On the other hand, back reflection occurs which partially increases the intensity of the primary beam.

Both effects are taken into account in the *dynamical theory*. This theory requires considerably more mathematics: The exact treatment of the passage of an electron wave through crystals requires the solution of the Schrödinger equation (quantum mechanics).

The effects a) and b) play a role in those cases where the reflected intensities are not small compared with the primary beam intensity. The

[1] As was mentioned in Chapter 1, the expression "reflected" has become common usage for those rays diffracted with the Bragg angle θ, although diffraction is merely a special case of reflection (which is possible at all angles) on individual lattice planes.

[2] The treatment in Sections 4.1–4.5.2 partially follows a paper by Alexander [3].

intensity of electron beam interferences is 10^6–10^7 times stronger than those of x-ray interferences (see Table 3.1). Consequently, the kinematical theory usually is sufficient for x-ray diffraction phenomena. An adequate treatment of electron diffraction, on the other hand, often requires the dynamical theory. The *kinematical theory,* however, gives a good *qualitative view* of the phenomena which occur during electron irradiation of crystalline samples. This is particularly true if lattice defects are present, a case of great practical importance. Therefore the most significant contrast effects explainable with the kinematical theory will be discussed in this chapter. An analogous treatment with the dynamical theory is not possible within the framework of this short introduction.

4.2. Derivation of the Basic Equation of the Kinematical Theory

After the introductory discussion of Section 4.1, the calculation of the electron intensity scattered at a crystal lattice is to be treated more precisely in this section.

Assume that a crystal K *is given* having arbitrary shape and finite size and being in an arbitrary orientation (Fig. 4.2) with respect to a plane monochromatic electron wave. The wave direction is defined by $\mathbf{k}_0$ with $|\mathbf{k}_0| = 1/\lambda$. $\mathbf{k}_0$ is the primary beam of intensity I_0. (In the previous chapters it has also been designated with $\mathbf{P}$.) The diameter of the beam is larger than the crystal diameter (i.e., the electrons "envelop" the crystal).

In the arbitrary spatial directions $\mathbf{k}$, *one wishes to determine* the scattering intensity $\mathbf{I}$ as a function of the shape and orientation of the crystal, i.e., $I = I(\mathbf{k},K)$, where I is proportional to A^2.

To solve this problem, one considers that all lattice atoms of the real crystal K are located at the end points of the lattice vectors $\mathbf{r}_n$ [given by Eq. (3.3)] of the real lattice. In Fig. 4.1 an arbitrary atom O has been chosen as the origin of the coordinate system, and the lattice vector $\mathbf{r}$ points to another atom P. The first step in the solution to the problem is the calculation of the phase difference between the partial waves scattered from these two atoms in the direction $\mathbf{k}$. This phase difference is based on the path difference of the two waves according to the geometry of Fig. 4.1. This path difference is $y - x$ where $y = r \cos \alpha$, $x = r \cos \beta$, and $r = |\mathbf{r}|$. The cosine functions can be defined by the *scalar product*

$$\cos \alpha = \mathbf{k}\,\mathbf{r}/(|\mathbf{k}|\,|\mathbf{r}|), \qquad \cos \beta = \mathbf{k}_0\,\mathbf{r}/(|\mathbf{k}_0|\,|\mathbf{r}|).$$

Thus

$$y = r\,\frac{\mathbf{k}\,\mathbf{r}}{|\mathbf{k}|\,|\mathbf{r}|} = \lambda\,\mathbf{k}\,\mathbf{r},$$

since $1/|\mathbf{k}| = 1/|\mathbf{k}_0| = \lambda$. The analogous equation is $x = \lambda\,\mathbf{k}_0\,\mathbf{r}$. Consequently, $y - x = \lambda(\mathbf{k} - \mathbf{k}_0)\mathbf{r}$. Since the following relation is valid:

$$\frac{\text{path difference } (y - x)}{\lambda} = \frac{\text{phase difference } \Delta\varphi}{2\pi},$$

the phase difference is $\Delta\varphi = (2\pi/\lambda)(y - x)$, or

$$\boxed{\Delta\varphi = 2\pi(\mathbf{k} - \mathbf{k}_0)\,\mathbf{r}} \;. \tag{4.2}$$

Therefore the generalized statement can be made that the phase difference between the origin and the nth atom located at $\mathbf{r}_n$ equals $2\pi(\mathbf{k} - \mathbf{k}_0)\,\mathbf{r}_n$.

The waves emanating from the n atoms have a uniform wavelength λ but different phases φ_n and eventually different amplitudes f_n. They are summed according to Eq. (4.1) to yield the total amplitude A.

By substituting Eq. (4.2) into (4.1), the basic equation of the kinematical theory is obtained:

$$\boxed{A\,(\mathbf{k}) = \sum_n f_n \exp[2\pi i(\mathbf{k} - \mathbf{k}_0)\mathbf{r}_n]} \;. \tag{4.3}$$

(Sometimes the exponent is written with a negative sign; both forms of the equation are equivalent.) If the crystal is subdivided into elementary regions (volume V_0), and the scattering factor of one of them is called $f(\mathbf{r})$, the last equation can be written as an integral:

$$A\,(\mathbf{k},K) = \frac{1}{V_0}\int_K f(\mathbf{r})\exp[2\pi i(\mathbf{k} - \mathbf{k}_0)\mathbf{r}_n\,dV. \tag{4.4}$$

This solution gives the scattering amplitude A as a function of the following *independent variables:*

- *shape* of crystal K,
- *orientation* of crystal (position of $\mathbf{r}_n$ with respect to $\mathbf{k}_0$),
- *scattering direction* $\mathbf{k}$, where an observer or a measuring instrument would register the intensity scattered in this direction.

In the following sections, the term dV in the integral is sometimes replaced by $d\mathbf{r}$, which stands for integration over the whole crystal. Also, the dependence of A on the shape and orientation of the crystal was designated by the letter K. This letter is not written in subsequent sections, although the dependence on K should be kept in mind.

4.3. Ewald Sphere

Up to now, the general case was treated where the crystal is composed of different types of atoms with their corresponding atomic scattering amplitudes. However, in many applications to be discussed, a simplification is appropriate, namely, that only one type of atom is present, as in pure metals. In that case, the $f(\mathbf{r})$ in Eq. (4.4) are a constant with respect to the integration, and with 2θ = angle $(\mathbf{k}, \mathbf{k}_0)$ one obtains

$$\mathbf{A}\ (\mathbf{k}) \sim \underbrace{f(\theta)}_{\substack{\text{atomic} \\ \text{contribution}}} \underbrace{\int_K \exp[2\pi i(\mathbf{k} - \mathbf{k}_0)\mathbf{r}\ dV}_{\substack{\text{lattice} \\ \text{contribution}}} \ . \qquad (4.5)$$

atomic contribution to the scattering amplitude

In this case, the atomic contribution and the lattice contribution are separate factors.

The form of the last three equations shows immediately that the *absolute maximum* of A occurs if, and only if, the expression $(\mathbf{k} - \mathbf{k}_0)\mathbf{r}$ is an *integer* for all positions $\mathbf{r}$. In that case the exponential function has the value 1 and $A = Nf(\theta)$, where N is the number of atoms in the crystal.

What is the meaning of this condition for the maximum of A? $(\mathbf{k} - \mathbf{k}_0)\mathbf{r}$ is an *integer when* $(\mathbf{k} - \mathbf{k}_0)$ *is a vector of the reciprocal lattice*. This vector is called $\mathbf{g}$ (see Chapter 3). The proof for this proposition is found in Eqs. (3.3) and (3.4). The multiplication of an arbitrary vector $\mathbf{r}$ in the real lattice [Eq. (3.3)] with an arbitrary vector $\mathbf{g}$ in the reciprocal lattice [Eq. (3.4)] results in the equation [use the system of Eq. (3.5)]:

$$\mathbf{rg} = mh + nk + pl = \text{integer} \qquad (4.6)$$

because all numbers m, n, p, h, k, l are integers.

The question now arises as to the physical significance of this statement. It can be interpreted visually with the geometrical image of the so-called *Ewald sphere* (also called reflecting sphere): First the primary radiation vector $\mathbf{k}_0$ is drawn which by definition has the length $1/\lambda$ (Fig. 4.3). The end point of this vector is taken as the Origin O^* of the reciprocal lattice of K. In the figure, this lattice is indicated by several spots. Around the starting point of $\mathbf{k}_0$ a sphere is drawn whose radius equals the length of this vector, namely $1/\lambda$, i.e., the sphere passes through O^*. This sphere is the Ewald sphere. A maximum of A occurs when by chance a reciprocal lattice point $\mathbf{g}$ (abbreviated rel-point, see Chapter 3) falls exactly on the sphere surface. In that case $\mathbf{k} - \mathbf{k}_0 = \mathbf{g}$, i.e., the difference is a vector in the reciprocal lattice, as can be seen in Fig. 4.3. Furthermore, it can be seen that the lattice planes (hkl)[3] associated with g are in *Bragg reflection* position with $1/d = |g|$; the lattice planes are normal to $\mathbf{g}$ and Fig.

---

[3] Although Fig. 4.3 is a representation in reciprocal space, the lattice planes (hkl) belonging to real space (and which are normal to $\mathbf{g}$) have been drawn for the sake of clarity.

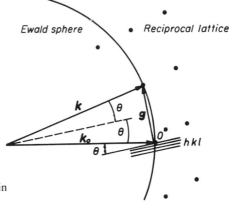

Fig. 4.3. Definition of the Ewald sphere in
the reciprocal lattice.

4.3 shows that

$$\sin \theta = \frac{g/2}{1/\lambda} = \frac{\lambda g}{2} = \frac{\lambda}{2d} \quad \text{or} \quad \boxed{\lambda = 2d \sin \theta} \,, \tag{4.7}$$

i.e., the Bragg condition. In the nomenclature of the reciprocal lattice and
with the vectors $\mathbf{k}$ and $\mathbf{k}_0$ the Bragg condition is

$$\boxed{\mathbf{k} - \mathbf{k}_0 = \mathbf{g}} \,. \tag{4.8}$$

This independent derivation of the Bragg equation from the basic equation (4.3) or (4.5) obviously is a necessary condition for its correctness,
because the prerequisites for this derivation and for the more simple one
given in Chapter 1 are the same: namely, maximum intensity caused by
superposition of several waves with phase differences of integral multiples
of λ.

Consequently, the Ewald sphere, for electron diffraction as well as for x-ray diffraction,
has the following graphic meaning: It provides immediately those directions $\mathbf{k}$ in reciprocal
space which are coordinated with secondary rays causing reflections in real space. These
reflections are the intersections of the sphere surface with the reciprocal lattice points.
Although reciprocal and real space, having different dimensions, should be kept strictly
separate, yet in certain respects one can view the Ewald sphere as being identical with the
photographic plate since, with respect to the angles, both are the locus of the reflections.
For this reason the Ewald sphere is also called the reflecting sphere. Also, the angles, such
as θ and 2θ, are the same in both spaces. Distances, however, have to be converted with
the "scaling factor" λL [nm $\times$ mm], as derived in Section 3.3.

In Section 3.3, Theorem 2 stated that the diffraction pattern corresponds to a *nearly plane
section* through the reciprocal lattice. This theorem can now be worded more precisely by
stating that the diffraction pattern corresponds exactly to the *surface of the Ewald sphere*.
Its curvature which is a measure of the approximation embodied in Theorem 2, will be made
clearer by a calculation in Section 4.5.2 (commentary to Fig. 4.10).

4.4. Amplitude–Phase Diagram

Before any application examples for the basic equation are discussed it will be useful to give a clear description of the principles derived in Section 4.3.

The *theory of vibrations* furnishes a simple method for obtaining the resultant vibration from a number of individual vibrations of different phase: Each vibration is represented by a vector whose length is proportional to the amplitude, while the angle between the vectors indicates the phase angle between the individual vibrations. The resultant vibration is obtained by vector addition. (This can easily be proved by describing sine vibrations as the projection of a circular motion.) The diagram constructed with the help of amplitudes and phases is called the "amplitude–phase diagram."

This method is valid also for the amplitudes of *waves* with the wave vector **k** as shown in Fig. 4.1. In this context, the time dependence of wave elongation E is of no interest. However, if at a fixed time, E is plotted as a function of its location (space coordinate x), sine curves are obtained. The wave amplitudes are drawn as vectors at the maxima of this sine curve. In Fig. 4.4a, the amplitude of the individual wave is designated A_0 and drawn as a symbolical arrow. If two waves with phase difference zero or $2\pi = \lambda$ are superimposed, the resultant wave (dashed line in Fig. 4.4b) has the amplitude $A = 2A_0$. If the phase difference is π, the two wave cancel each other because the two vectors A_0 have opposite directions (Fig. 4.4c). The construction of the amplitude A for the case of an arbitrary phase difference is shown in Fig. 4.4d: A is obtained

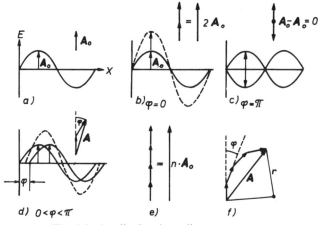

Fig. 4.4. Amplitude–phase diagram (see text).

as the vector sum of the amplitude–phase diagram (APD) from the two vectors A_0 rotated one versus the other by the angle φ.

If not two, but n waves are superimposed, all having the same phase difference, the usefulness of the APD is particularly evident. If all waves have the same phase, the resultant amplitude is $A = nA_0$ (Fig. 4.4e). This case of the maximum possible amplitude is the repeatedly mentioned case when the condition of the Bragg equation is satisfied ("exact Bragg case"). If, on the other hand, a finite phase difference φ exists, the APD according to Fig. 4.4f is a circular arc, or possibly a full circle, which may even have been run several times. The calculation of the radius r of this circle will be given in Section 4.5.2.

In the meantime, it is already possible to treat qualitatively a very important case, namely the so-called *fringe contrast* originating at *wedge-shaped crystals*. (This will be done with the help of the APD within the framework of the kinematical theory.) These fringes caused by interference are also termed wedge interferences. The case is important because electrolytically thinned foils always have more or less wedge-shaped edges. However, the reasoning applied to wedges is also valid for all slanted interfaces in the interior of crystal foils, such as grain boundaries, or, in a modified form, for stacking faults (Section 4.6.2). Examples of wedge interferences can be seen, e.g., in Fig. 1.10 (fringe contrast at grain boundaries), Fig. 4.13 (at the wedge-shaped sample edge), or in Fig. 4.14 (at twin boundaries).

In Fig. 4.5 the wedge-shaped crystal is drawn in cross section. Two cases have to be distinguished:

a) If the orientation of the crystal with respect to the primary beam happens to correspond exactly to a Bragg position, the APD for the scattering amplitude A, according to Fig. 4.4e, consists of a vector A whose length is proportional to the wedge thickness. The scattering intensity I_r

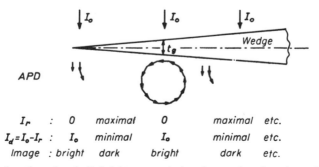

Fig. 4.5. Explanation of periodic brightness variations in a wedge-shaped sample by means of the APD (modified after Alexander [3]).

is proportional to the square of **A**. Consequently, if the crystal is in exact reflection position for any one lattice plane (reciprocal lattice vector **g**), the intensity removed from the primary beam increases with the thickness of the crystal. In this case the image is getting darker with increasing sample thickness ("gray wedge").

b) If, on the other hand, the crystal is not in exact Bragg position, but rotated by a small angle away from this orientation, the secondary waves from neighboring atoms will have phase differences φ as shown in Fig. 4.4f. The APD becomes an *arc* whose length increases proportionally with the sample thickness (Fig. 4.5). At its edge, the wedge image is bright because in the beam direction only few atoms are present in a plane of approximately reflecting position. With increasing wedge thickness the APD becomes a *semicircle: $I_r \sim A^2$* is a maximum and $I_d = I_0 - I_r$ is a minimum, i.e., the image shows a *dark fringe*. (It is understood that the term "image" means the bright field image.) If the thickness increases further, obviously the resultant vector **A** decreases again since the chord of a circular arc (Fig. 4.4f) becomes shorter when the arc becomes longer than a semicircle. When the circle closes completely, I_r again equals zero, i.e., the transmitted intensity I_d equals the primary intensity I_0. This case is indicated in the middle of Fig. 4.5. As a consequence of the complete transmittance the image shows a *bright fringe*. With further increasing sample thickness this sequence is repeated. Thus the system of parallel bright and dark fringes is formed which is termed *fringe contrast*. They are also called "thickness fringes."

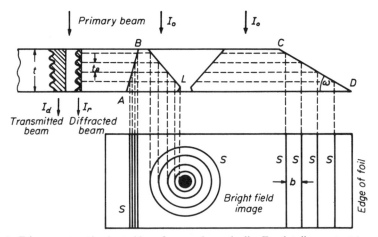

Fig. 4.6. Fringe contrast in slanted interfaces, schematically. For details, see text (according to Thomas *et al.* [2]).

Thus a periodic *oscillation* of I_r as well as I_d occurs as a function of the sample thickness or the depth t in the foil. This oscillation is sketched at the left of Fig. 4.6. The upper part of the figure shows the foil in cross section. The period of brightness oscillation t_g depends on **g** and φ; it will be calculated in Section 4.5. It can be visualized, however, that t_g decreases as the deviation φ increases, since the radius r in Fig. 4.4f decreases as φ increases. t_g is also called the *extinction distance* (here within the framework of the kinematical theory). Figure 4.6 shows schematically a steep grain or twin boundary $A - B$, a hole L, and the wedge-shaped sample edge $C - D$. In the lower part of the figure the fringes S, as explained above, are shown as they appear in the bright field image. Those fringes are obtained—this should always be kept in mind—by *projection of all sample details* onto the image plane. According to Fig. 4.6 the following relation exists between t_g and the fringe distance b: tan ω = t_g/b (ω is the wedge angle).

A prerequisite for this derivation of contrast is that the *absorption* in the sample is negligible, compared with the diffracted intensity I_r. According to Chapter 1, this prerequisite is usually satisfied. On the other hand, if in addition to diffraction contrast, absorption, which increases with thickness, is present, nothing fundamental is changed in the derivation above. A darkness gradually increasing with thickness is merely superimposed on the fringe contrast. However, contrast changes due to the limitations of the kinematical theory will be discussed briefly in Section 4.5.3.

4.5. Application of the Basic Equation to Ideal Crystals

4.5.1. Deviation Parameter s and its Experimental Determination

It is practical to use the basic equation with the help of the Ewald sphere in the following way: The primary beam $\mathbf{k}_0$ is assumed to be given and one wants to determine the resultant amplitudes A in the independently variable directions $\mathbf{k}$. The crystal K or its reciprocal lattice (K^*) may be another independent variable. The crystal can assume any arbitrary position, i.e., orientation with respect to $\mathbf{k}_0$, by rotation in any direction around the crystal origin O^*. If a rel-point **g** does *not* lie exactly on the Ewald sphere at the end point of the vector **k**, as in Fig. 4.3 [satisfying Eq. (4.8)], but near the sphere, one expects intuitively that the scattering amplitude $A(\mathbf{k})$ is smaller than in the case where the Bragg equation is satisfied. $A(\mathbf{k})$ will increase as the distance of the rel-point **g** from the end point of vector **k** decreases. This distance is designated with **s**; it is an important parameter. The vector **s** is a measure of the *deviation from that condition which exactly satisfies the Bragg equation*, i.e., where $\mathbf{k}_0 + \mathbf{g} = \mathbf{k}$. **s** is a vector in reciprocal space, as are **g**, **k**, and $\mathbf{k}_0$; they have the dimension [nm^{-1}].

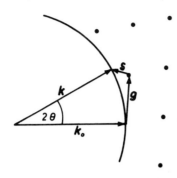

Fig. 4.7. Definition of the vector **s** when **k** is given. **s** is a measure of the deviation from the exact satisfaction of the Bragg equation, where $\mathbf{k}_0 + \mathbf{g} = \mathbf{k}$ would be true.

According to Fig. 4.7:

$$\boxed{\mathbf{k} - \mathbf{k}_0 = \mathbf{g} + \mathbf{s}} \tag{4.9}$$

$s = 0$ corresponds to the exact satisfaction of the Bragg condition; in the diffraction pattern the respective reflection has maximum intensity. If the Ewald sphere does not intersect another rel-point, $s = 0$ is identical with the so-called *two-beam case* of Section 3.11.

An arbitrary $s \neq 0$ (the direction of **s** does not have to be parallel to $\mathbf{k}_0$) corresponds to an independently variable given direction **k** for which the scattering amplitude $A(\mathbf{k})$ is to be determined. One substitutes Eq. (4.9) into Eq. (4.5). Considering the fact that $\mathbf{g} \cdot \mathbf{r}$ always is an integer, the scattering amplitude can also be written as

$$\boxed{A(\mathbf{s}) \sim f(\theta) \int_K \exp[2\pi i \mathbf{s} \cdot \mathbf{r}]\, dV} \tag{4.10}$$

Experimental determination of the deviation parameter s

The experimental determination of **s** is possible in all those cases where the diffraction pattern contains *Kikuchi lines* (see Section 3.11). $s = 0$ corresponds to the exact Bragg case (two-beam case) for the interference g in question. According to Section 3.11, the correlated bright Kikuchi line in that case must pass *exactly through* the Bragg reflection **g** (which has maximum brightness), and the dark line must pass through the primary beam (zero-order reflection). This is the case for the $5\bar{1}\bar{1}$ reflection in Fig. 3.17. If, however, the crystal is tilted by a small angle ϵ away from the Bragg position, the system of Kikuchi lines undergoes a parallel shift along the vector **a**, as explained in Section 3.11. **a** is normal to the tilt axis. Figure 4.8A shows the situation for $s = 0$, corresponding to $\epsilon = 0$, and for $s \neq 0$ (after v. Heimendahl [4]). Point P represents the end point of the rel-vector **g** for the case $s = 0$, when the Kikuchi line L_g passes exactly *through* the Bragg reflection, and the reflected beam k_g, here identical for both the Kikuchi line *and* the Bragg reflection, intersects the Ewald sphere in P. If the sample is tilted by the angle ϵ, point P moves to P_1 because of the simultaneous rotation of the reciprocal lattice around O^*; the distance $P - P_1$ equals s. The angle between k_g and the reflected beam k_1 leading to the bright Kikuchi line L_1 also is ϵ. The latter fact is based on the mechanism of Kikuchi line formation. Since ϵ is a small angle, it can be seen in the drawing for reciprocal space (Fig. 4.8A) that

$$\epsilon = s/g. \tag{4.11}$$

Since, on the other hand, according to Section 3.11, $\epsilon = a/L$ (L is the camera length) in real space, it follows:

$$s = ag/L. \tag{4.12}$$

The *sign of s* is defined by *convention:* s is positive if P_1 lies in the *interior* of the Ewald sphere, s is negative for the reverse case. Thus, for $s < 0$, the bright Kikuchi line L_1 lies between the Bragg reflection and the primary beam, for $s > 0$, the line is "outside" the Bragg reflection.

The aluminum diffraction pattern, Fig. 3.17, which already has been analyzed extensively in Section 3.11, may serve as an example for the determination of s. The bright Kikuchi line L_1 of the $(15\bar{1})$ reflection is shifted by $a = 2$ mm to the "outside," i.e., s is positive. L_1 is normal to $g(15\bar{1})$. The camera length L can be calculated from the diffraction distance measured on the diffraction pattern photograph: $\lambda L = 2.02$ [nm × mm] and $\lambda(100$ kV$) = 0.0037$ nm; therefore $L = 545$ mm. Since $g(15\bar{1}) = 1/d(15\bar{1}) = 12.85$ [nm^{-1}], one obtains from Eq. (4.12):

$$s(15\bar{1}) = \frac{2 \times 12.85}{545} = 0.047 \ [\text{nm}^{-1}].$$

The accuracy of the s determination depends on the sharpness of the Kikuchi lines. As mentioned in Section 3.11, the sharpness depends on the local crystal perfection. The latter is different for the different lattice planes in the example of Fig. 3.17.

s in the symmetrical (Laue) case

Figure 4.8A is valid for *an arbitrary sample orientation* in the neighborhood of a Bragg position. Figure 4.8B shows the *special sample positions* of the Laue case and the Bragg case defined in Section 3.11, including the positions of all pertinent **k** vectors and the appearance of the diffraction pattern and the Kikuchi lines. In anticipation of Section 4.5.2,

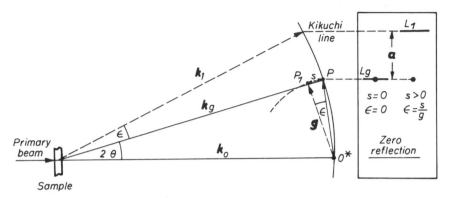

Fig. 4.8A. Location of those secondary beams leading to Kikuchi lines in reciprocal space (left), and corresponding diffraction pattern (right). The Kikuchi lines L_g and L_1 correspond to the cases $s = 0$ and $s > 0$, respectively (after v. Heimendahl [4]). General case.

the figure makes use of the fact that in practice, when thin foils are irradiated, s is always normal to g. As a consequence, the reflecting lattice points in the symmetrical case ($\mathbf{g} \perp \mathbf{k}_0$) are all *outside* the Ewald sphere. Therefore, according to the *convention, s is always negative for the symmetrical case*. From Fig. 4.8B it can be seen:

$$\text{Symmetrical case:} \boxed{s = -\theta g} \quad .$$

For the Al(200) diffraction spot $d = 0.2025$ nm, $g = 1/d \approx 5$ nm^{-1}, and θ is calculated from Eq. (1.6): With $\lambda = 0.0037$ nm (100 kV), $\theta = 0.915 \times 10^{-2}$ or approximately $\frac{1}{100}$ ($\approx \frac{1}{2}°$). With these data $s = -0.05$ nm^{-1}. For the (600) diffraction spot, s is nine times that amount.

Further methods for the determination of s by means of interference contours will be discussed in Sections 4.5.4, 4.5.5 [Eq. (4.21)], and 4.6.1.3.

4.5.2. Crystal in the Shape of a Brick-Shaped Parallelepiped

For the applications to follow, the atomic scattering amplitude $f(\theta)$ is sometimes omitted in the equations for $A(\mathbf{s})$ or $A(\mathbf{k})$, since in most cases $f(\theta)$ plays a minor role, compared to the lattice factor, Eq. (4.5).

A simple and important example is the calculation of the scattering amplitude of a crystal which has the shape of a parallelepiped with dimensions a, b, and c. In an orthogonal coordinate system, the components of $\mathbf{s}$ are called s_x, s_y, and s_z, and the atom positions s, y, and z. Equation

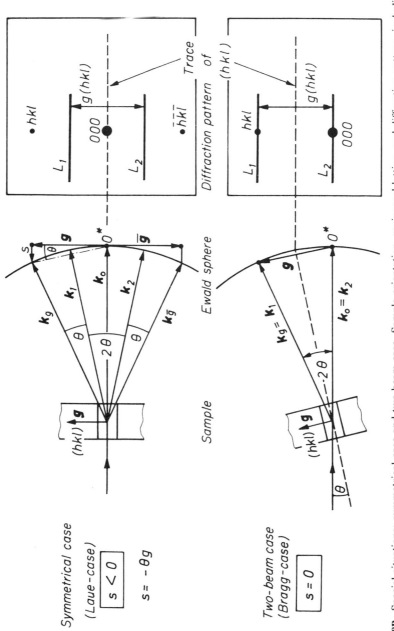

Fig. 4.8B. Special situations: symmetrical case and two-beam case. Sample orientation, reciprocal lattice, and diffraction pattern, including Kikuchi lines, are shown. $\mathbf{k}_1$ and $\mathbf{k}_2$ are those secondary beams leading to the Kikuchi lines L_1 and L_2 (modified after Thomas [4a]).

(4.10) then becomes

$$A(\mathbf{s}) \sim \int_{-a/2}^{a/2} \int_{-b/2}^{b/2} \int_{-c/2}^{c/2} \exp[2\pi i\,(s_x x + s_y y + s_z z)]\,dx\,dy\,dz.$$

This integral can be solved by elementary means, since, e.g., the following holds:

$$\int_{-a/2}^{a/2} \exp[2\pi i s_x x]\,dx = \frac{1}{2\pi i s_x}\,(\exp[\pi i s_x a] - \exp[-\pi i s_x a]) = \frac{2i\sin \pi s_x a}{2\pi i s_x}.$$

Consequently

$$A(\mathbf{s}) \sim \frac{\sin \pi s_x a}{\pi s_x}\,\frac{\sin \pi s_y b}{\pi s_y}\,\frac{\sin \pi s_z c}{\pi s_z}. \qquad (4.13)$$

First let us view a, b, and c as given quantities. If one asks for the dependence of the *amplitude* of $\mathbf{s}$, one obtains, e.g., A as a function of s_x as shown in the curve of Fig. 4.9a. By squaring that curve one obtains the curve of Fig. 4.9b which is proportional to the *intensity*. This curve has a primary maximum whose width is inversely proportional to the length a of the crystal in the corresponding direction, and whose height is proportional to the square of the same length a. Nearly the whole reflected intensity is contained in this primary maximum. Therefore the curve is easily interpreted as follows: For *very large a*, the primary max-

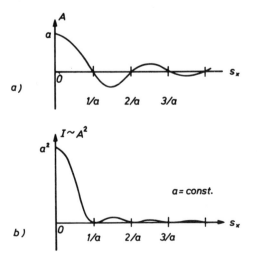

Fig. 4.9. a) Amplitude and b) intensity as a function of s_x.

imum is very high and steep, for $a = \infty$ it is a δ-function. If the whole crystal were "infinitely large," this interpretation would hold for all three spatial directions, and intensity would be present only for $s = 0$: In this case the "region capable of reflection" in the reciprocal lattice would in fact be only the ("infinitely small") rel-point g itself. Actually, the assumptions of the kinematical theory would be exceeded for this degenerated case. However, even within the framework of the dynamical theory, the statement would be qualitatively correct.

If, on the other hand, a, for example, is assumed to be finite, and $b = c = \infty$, the sample is a *thin foil* if $a\|k_0$.[4] In that case, noticeable intensity I_r [according to Eq. (4.13)]

$$I_r \sim \left(\frac{\sin \pi s_x a}{\pi s_x} \right)^2 \qquad (4.14)$$

will be present also for small deviations from the exact Bragg orientation. This will occur for all those vectors k which correspond to an s-vector parallel to k_0, i.e., where s has only an s_x component. *In this context one has to visualize the reciprocal lattice point g no longer as a mathematically defined point, but as the center of a region capable of reflection,* corresponding to the intensity distribution given by Fig. 4.9b and Eq. (4.14).

Thus we have proved the important

Theorem 4. For a thin foil the rel-points are degenerated into rods (streaks) normal to the plane of the foil, and having a length of approximately $2/a$. Reflection occurs for all those locations defined by the vector k on the Ewald sphere, where the latter intersects a rel-rod, Fig. 4.10.

When the length of the rel-rods was stated to be $2/a$, the justified assumption was made that the secondary maxima in Fig. 4.9b are negligible. A check with Eq. (4.14) shows that for $s_x = 3/(2a)$, i.e., for the first secondary maximum, the intensity I_r is only $[2/(3\pi)]^2 a^2$, or 4.5% of the intensity in the primary maximum.

Appropriate generalization of Theorem 4 leads to

Theorem 5. For a brick-shaped crystal sample with dimensions a, b, c, the reciprocal lattice region capable of reflection is a "brick" with dimensions $\Delta s = 2[1/a, 1/b, 1/c]$ and the rel-point g as its center. *The region*

---

[4] a is the thickness of the foil, otherwise called t in this volume (Sections 1.6, 3.8, and 3.9). In this chapter, the letter a is retained to help the reader keep in mind the origin of the dimension, namely, the parallelepiped with edge lengths a, b, and c.

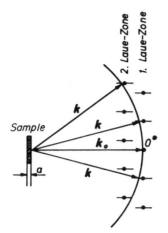

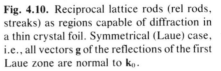

Fig. 4.10. Reciprocal lattice rods (rel rods, streaks) as regions capable of diffraction in a thin crystal foil. Symmetrical (Laue) case, i.e., all vectors **g** of the reflections of the first Laue zone are normal to $\mathbf{k}_0$.

capable of reflection is always particularly large in that direction in which the dimension of the sample is particularly small.[5]

The preceding two theorems have far-reaching consequences. With the help of Theorem 4 and Fig. 4.10, one can now understand the fact, stated without proof in Chapter 3, that multispot diffraction patterns exist at all. If the Bragg equation were strictly satisfied, one would expect patterns with only a single diffraction spot. In reality, however, the Ewald sphere intersects not only one rel-point (corresponding to the exact Bragg case), but several of the expanded rel-rods. Often these belong to the same *reciprocal lattice plane*, also called a *Laue zone*, because the lattice planes defined by the rel-rods are planes of one zone. The Ewald sphere in Fig. 4.10 intersects three rel-rods of the rel-plane, called "1st Laue zone," and one rel-rod of the "2nd Laue zone." (In addition, small *distortions* of the crystalline sample will also enlarge the regions capable of reflection.)

Figure 4.10 has been sketched for the symmetrical case where all vectors **g** of the 1st Laue zone are normal to $\mathbf{k}_0$. In reality, the curvature of the Ewald sphere is much smaller (its radius larger) than can be shown in a drawing. A quantitative example, namely, the (200) reflection of aluminum, may serve to visualize the size relationships. The lattice spacing is $d \approx 0.2$ nm, consequently $g \approx 5$ nm^{-1} for the first reflection in the 1st Laue zone in Fig. 4.10. The distance of the sphere surface from this point equals s. For this example, which has been calculated at the end of Section 4.5.1, $s \approx -0.05$ nm^{-1}. Thus the distance between the two surfaces: the sphere and the plane tangent to the sphere at the origin O^* is only

[5] Theorems 4 and 5 are equivalent to the corresponding x-ray diffraction phenomenon, namely broadening of diffraction spots or lines because of small size of the diffracting particles.

$0.05/5 = 1/100$ of that vector **g** which corresponds to the first reflection. For the second reflection (400), this distance is $g/50$ because $\theta(400) \approx 1/50$,

Theorem 4 also explains why orientation determinations by means of diffraction spot patterns are so inaccurate (about $\pm 5\%$ as mentioned in Chapter 3.) Imagine the crystal (in Fig. 4.10) together with its fixed reciprocal lattice slightly rotated around the latter's origin O^*. Because of the length of the rel-rods, the diffraction pattern (the intersections of all rel-rods with the Ewald sphere) possibly may not change within a certain angular range of rotation. In first approximation, the diffraction spots will not change position, only intensity, corresponding to the intensity curve in Fig. 4.9b, as plotted along a rel-rod. (For quantitative investigations of this problem see Laird *et al.* [21].) Only when, with continued rotation of the crystal around O^*, the Ewald sphere intersects a new Laue zone, new diffraction spots will appear at their proper new locations. This behavior of spot diffraction patterns during sample tilting is completely different from that of Kikuchi lines (Section 3.11), which shift continuously with sample tilting. Both phenomena can be observed experimentally when a sample is tilted in the EM with the latter switched to diffraction.

Finally, there are possibilities, based on detailed analyses of the conditions described, to *increase* the accuracy of orientation determinations by means of spot patterns. These will be treated separately in Section 4.12.

If the periodicity distance $1/a$ between the maxima in Fig. 4.9b could be measured experimentally, the *determination of the foil thickness a* would be possible (see Section 4.5.5).

In Fig. 4.11, three sketches in perspective demonstrate the more important applications of Theorem 5. For a sample which is "large" in all three dimensions, only the *rel-points* themselves are "capable of reflec-

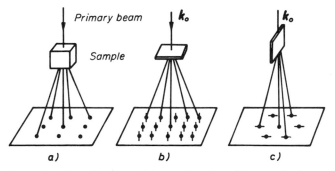

Fig. 4.11. Shape of reciprocal lattice points as a function of the sample shape: a) ("thick") sample extended in all three dimensions, b) and c) thin foils or crystals normal or parallel, respectively, to the primary beam.

tion'' (Fig. 4.11a). However, it should be pointed out again that this case cannot be treated adequately with the kinematical theory. Nevertheless, the statement of Fig. 4.11 is qualitatively correct in the general case. For the common case of a thin foil normal to k_0 (Fig. 4.11b), the *rel-rods* are normal to the plane of the diffraction pattern which, consequently, is a complete spot pattern. If the sample contains a thin plate-shaped object, as, for example, a decomposition zone or a thin platelet precipitate parallel to k_0, its reciprocal lattice will degenerate to rods *parallel to the image plane* (Fig. 4.11c). Such ''streaks'' of relatively high intensity are found, for example, in the diffraction patterns of plate-shaped GP zones in the precipitation hardenable Al–Cu alloys. An example is shown in Fig. 4.33 to be discussed later.

Up to now, the dimensions a, b, c in Eqs. (4.13) or (4.14) were kept constant and s was treated as an independent variable. Another possibility is to treat s (i.e., the crystal orientation) as a constant and vary one of the dimensions, e.g., the crystal thickness a, in the direction of the electron beam. In that case, Eq. (4.14) yields a simple $\sin^2$ dependence for the intensity I_r, with periodical zero values spaced $1/s$ (Fig. 4.12 top). Such a continuously increasing thickness is identical with a *wedge-shaped crystal* which has been treated qualitatively in Section 4.4 as an example for the application of the amplitude–phase diagram. In that treatment, the oscillations of the intensity I_r were derived graphically. They had the depth periodicity t_g defined as the extinction distance. These oscillations now are confirmed quantitatively. For t_g (Fig. 4.12), the following relation holds:

$$t_g = 1/s, \tag{4.15}$$

where the vector s defines the distance to that adjacent Bragg reflection g whose proximity to the Ewald sphere causes the fringe contrast according to Section 4.4.

In the lower part of Fig. 4.12 the wedge-shaped crystal is sketched

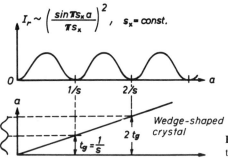

Fig. 4.12. Intensity as a function of sample thickness a (modified after Alexander [3]).

similar to Fig. 4.5 and 4.6. The a axis of the upper part of the figure is rotated 90° in the lower part.

Finally, with Eq. (4.13) the radius r of the circle in the amplitude–phase diagram (Section 4.4) can be determined as a function of s. According to Eq. (4.13) the *maximum possible amplitude* $A(\text{max})$ for a thin foil ($s_y = s_z = 0$) is $1/\pi s_x$, with the proportionality factor in Eq. (4.13) set equal to 1. This maximum amplitude equals the diameter $2r$ of the circle in the APD. Consequently, $2r = 1/\pi s$ or

$$r = 1/(2\pi s). \tag{4.16}$$

4.5.3. Extinction Distance

According to Eq. (4.15) the extinction distance t_g increases with decreasing deviation s; it would become infinite for $s = 0$, i.e., for the exact Bragg case. In the latter case, however, the conditions for the validity of the kinematical theory, except in very thin sample regions at the foil edge, are not satisfied, since strong reflection occurs for $s = 0$ (compare Table 4.1). $I_r \ll I_0$ is valid only for finite deviations s, while $I_r \approx I_0$ for $s \approx 0$ (two-beam case, e.g., diffraction pattern Fig. 4.13). Thus, in the latter case, usually the conditions of the dynamical theory are satisfied by definition. It can be shown that even if $s = 0$, a finite extinction distance exists. It is usually called ξ_g (to distinguish it from t_g) and can be calculated with the following formula:

$$\xi_g = \pi V (\cos \theta)/(\lambda F_g), \tag{4.17}$$

where V is the volume of the unit cell and F_g the *structure factor* of the reflection $g(hkl)$. This factor is calculated from the atomic scattering amplitudes f_n of the individual atoms in the unit cell by means of the following formula:

$$F_{hkl} = \sum_n f_n(\theta) \exp[2\pi i(hp_n + kq_n + lr_n)]. \tag{4.18}$$

p_n, q_n, r_n are the atom coordinates of the nth atom expressed as fractions of the lattice constants a, b, c of the unit cell. The *structure factor*, also called *structure amplitude*, is a measure of the scattering ability of the unit cell. Equation (4.18) applies the basic equation (4.3) to the unit cell.

The extinction distance ξ_g is one of the most important constants in the contrast theories, especially in the dynamical theory. To calculate it, the relativistically corrected values of F_g and λ have to be used. Table 4.2 lists some values for ξ_g; they can be calculated with Eq. (4.17) for each material (crystal lattice) and each reflection g.

If s is not zero, but not large enough to fulfil the prerequisites of the kinematical theory, an *effective extinction length* ξ_g^{eff} can be defined with

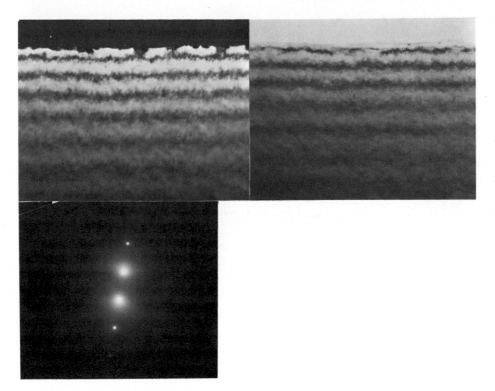

Fig. 4.13. Wedge fringes at the edge of an aluminum foil. Upper right: bright field image, upper left: dark field image taken with the strong (111) reflection in the diffraction pattern (lower left). In the latter the primary beam is the lower of the two reflections of nearly equal brightness (two-beam case), 20,000×.

Table 4.1 Comparison of kinematical and dynamical theory

Kinematical theory	Dynamical theory
$I_r \ll I_0$	I_r not very small compared to I_0
Realized in	
Thin sample areas, or large deviations from exact Bragg position ($s \neq 0$)	Thick sample areas, or orientation near exact Bragg position ($s \approx 0$)
Mathematical treatment	
Addition of scattering amplitudes of lattice atoms. (Only the phase differences caused by path differences are considered, as for the derivation of the Bragg equation.)	Quantum mechanics, i.e., solution of the Schrödinger equation including boundary conditions. (Consideration of extinction, interaction of multiply diffracted beams, etc.)

Table 4.2 Examples for extinction distances ξ_g[a,b]

hkl	Al	Cu	Au	MgO	hkl	Fe	Reflection	Mg	Co	Zn
111	55.6	24.2	15.9	272.6	110	27.0	$\bar{1}100$	150.9	46.7	55.3
200	67.3	28.1	17.9	46.1	200	39.5	$11\bar{2}0$	140.5	42.9	49.7
220	105.7	41.6	24.8	66.2	211	50.3	$\bar{2}200$	334.8	102.7	118.0
311	130.0	50.5	29.2	1179.7	220	60.6	$\bar{1}101$	100.1	30.6	35.1
222	137.7	53.5	30.7	85.2	310	71.2	$\bar{2}201$	201.8	62.0	70.4

[a] All measurements in nanometers.
[b] After Hirsch et al. [1] which contains a detailed table.

the formula:

$$\xi_g^{\text{eff}} = \frac{1}{(s^2 + 1/\xi_g^2)^{1/2}} . \tag{4.19}$$

This equation includes all cases discussed so far.

a) For $s = 0$, $\xi_g^{\text{eff}} = \xi_g$ (maximum value of ξ_g^{eff}, dynamical case).

b) For $s^2 \gg 1/\xi_g^2$, Eq. (4.19) becomes Eq. (4.15), i.e., $t_g = 1/s$ which is the kinematical limiting case.

c) Between those two limits there is an area of transition. To decide quantitatively whether a situation is "dynamical" or "kinematical", one compares s with $1/\xi_g$: Depending on whether $s = 0$, s is comparable to $1/\xi_g$ or $s \gg 1/\xi_g$, the dynamical case, the transition case, or the kinematical case, respectively, is present. The extinction distance is of practical significance because it furnishes a means for determining layer thickness. Examples will be discussed in Sections 4.5.4 and 4.5.5.

4.5.4. Application Examples: Wedge Fringes, Bend Contours

From Eq. (4.14)

$$I_r \sim \left(\frac{\sin \pi s a}{\pi s} \right)^2 ,$$

shown graphically in Figs. 4.9 and 4.12, it is evident that in principle two means exist for obtaining intensity oscillations and, as a consequence, fringe contrast bands: variation of the thickness a or of the deviation parameter s.

The first condition is realized by any wedge-shaped layer or by any boundary plane positioned obliquely in the foil (grain boundary, twin boundary, stacking fault).

In this context, the so-called two-dimensional boundary planes are not to be regarded as "crystal defects" (as, e.g., dislocations in Section 4.6); they merely separate two regions of "ideal crystals." At a grain boundary, for example, the grain whose orientation is close to $s = 0$ furnishes the fringes, while the other grain with perhaps $s \gg 0$ is without contrast. The case of stacking faults will be treated separately in Section 4.6.2.

s can be varied, i.e., deviation from the exact Bragg position can be achieved by tilting the sample in the EM. In addition, such deviation occurs also when the foil is *bent*. In that case, the orientation of the lattice plane g(hkl) with respect to the primary beam k_0 is continuously varied locally. During preparation, the delicate thin crystal foils often are slightly distorted. At those locations where the Bragg equation is exactly satisfied, a significant amount of intensity is removed from the primary beam by diffraction on the respective lattice planes (*hkl*). Consequently, the bright field image contains a dark fringe at the corresponding place (see Section 1.6.2). Such a fringe is also called bend contour, extinction contour, or interference fringe. These fringes should not be confused with the fringe contrast produced at wedge-shaped samples, as discussed in Section 4.4. The latter fringes, therefore, are better called wedge fringes. Table 4.3 gives an overview of these two fringe contrast phenomena and their characteristics.

Wedge fringes

The formation of the typical parallel lines of alternating bright and dark fringes has been theoretically derived in Sections 4.4 and 4.5.2. Examples

Table 4.3 Fringe contrast phenomena in homogeneous crystal foils

Cause	Variation of the orientation or deviation parameter, s	Variation of the layer thickness, a
Realized in	Bent foils	Wedge-shaped foils
Phenomenon observed (identical nomenclature for the contrast)	Bend contours, extinction contours, interference fringes, fringes of equal inclination	Wedge fringes, thickness fringes, fringes of equal thickness
Examples	Figs. 2.10, 4.15	Figs. 1.10, 4.13, 4.14
Differentiation criterion	In dark field, only one fringe is bright (that fringe which gave rise to the respective Bragg reflection)	In dark field (with relevant Bragg reflection) all fringes are bright
Position of fringes	In arbitrary directions	Parallel to the wedge-shaped foil edge or to the trace of the boundary plane (grain or twin boundary, stacking fault)
Determination of foil thickness (Section 4.5.5)	Method of Siems *et al.* [6]	Counting of fringes

are shown in Fig. 4.13 (foil edge of an Al foil), Fig. 4.14 (coherent twin boundaries), and Fig. 1.10 where the grain boundaries have the characteristic striations.

The diffraction pattern of Fig. 4.13 indicates that two-beam conditions exist. The dark field image taken with the strong (111) reflection (left) is juxtaposed to the bright field image (right). As required by the theory (compare I_d and I_r in Fig. 4.6), the minima and maxima in the two images are complementary, and *all* fringes are bright in the dark field image. The fringes are exactly parallel to the foil edge.

The sample in Fig. 4.14 is "TD nickel," a dispersion hardenable alloy of nickel with 2 vol% ThO_2. The ThO_2 particles, incorporated by powder metallurgical means, produce a material of high-temperature strength (rocket nozzles) up to temperatures close to the melting point of the nickel matrix. The particles are visible in the figure as nearly round objects, appearing black because of mass contrast. Their size is 10–200 nm, with an average value of 40 nm [5]. They considerably inhibit dislocation movement and, especially at high temperatures (around 1000° C), suppress recrystallization [5]. If recrystallization does occur, the *recrystallization twins,* typical for nickel, are *smaller* by more than an order of magnitude, compared to those in pure nickel. Because the ThO_2 particles form obstacles to growth, the twins are severely serrated. In Fig. 4.14 these twins are clearly seen, including their typical fringe contrast. The straight coherent twin boundaries (1) can be distinguished from the irregularly curved incoherent twin boundaries (2). For the twin boundary at (3), the contrast conditions for wedge fringes ($s \approx 0$) are not satisfied. Therefore the boundary shows no fringes, or these are not resolvable because of the magnitude of s.

Bend contours

Examples can be found in several figures, as in Fig. 2.10, at the locations marked S and M, and in Fig. 4.15.

In actual images, bend contours often occur in *pairs,* as, e.g., in the above-mentioned figures. This can be explained with the help of Fig. 4.16. In relation to the primary beam, the *foil is bent symmetrically at the respective sample location.* The figure shows a cross section of the foil with the relevant lattice planes normal to the plane of the paper. In Fig. 4.16a the foil, when viewed from above, has a cylindrically concave curvature ("trough"). At only two places the lattice planes (hkl) and $(\bar{h}k\bar{l})$ are in Bragg position. The corresponding rays leading to the respective diffraction spots are drawn as double arrows. The dark interference fringes generated in the bright field image are drawn in projection, ac-

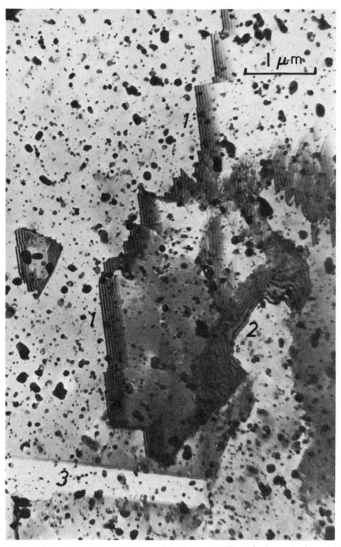

Fig. 4.14. TD nickel with coherent (1) and incoherent (2) twin boundaries. At (3) a twin boundary is out of contrast. The black spherical particles are ThO_2. Sample annealed after cold rolling (after v. Heimendahl [5]).

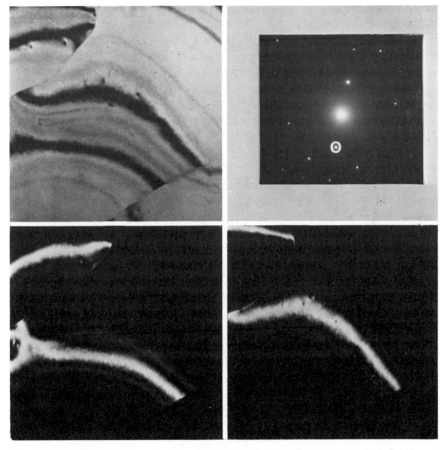

Fig. 4.15. Bend contours in pairs in an Al 99.99 foil, cold rolled 50%. The dark field image at lower right was made with the circled (200) reflection. Only *one* bend contour is in contrast. The dark field image at left was made with the opposite ($\bar{2}$00) reflection. It follows that the foil is locally bent as sketched in Fig. 4.16b, 40,000 ×.

cording to the mechanism of image formation (Section 1.6), under the cross section of the foil.

If the foil locally has a convex curvature, as viewed from above (Fig. 4.16b, "roof"), the rays leading to the diffraction spots go crosswise, and the contours (hkl) and $(\bar{h}\bar{k}\bar{l})$, which are in the same positions if the radius-of-curvature r is the same, exchange their indices. Thus, to determine whether case a) or case b) is present, one only has to take a dark field picture with the Bragg reflection (hkl) or $(\bar{h}\bar{k}\bar{l})$ and observe which of the two contours lights up brightly.

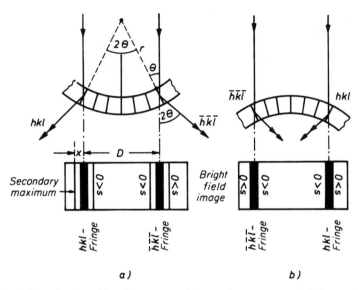

Fig. 4.16. Origin of pairs of bend contours with sample curvature of different sign. For details see text.

Figure 4.15 shows an example. From the correlation given in the caption, it is evident that in this case the foil is bent as a "roof."

An EM image also may contain *two* (or more) intersecting pairs of contours, e.g., Fig. 2.10 at one of the places marked S (right half of figure). In this case, the foil is not cylindrically curved, but has two mutually perpendicular radii of curvature, possibly of different magnitude. The method of analysis given above can obviously be expanded to include this three-dimensional case. One can then determine whether the foil is locally bent "dome-shaped," "bowl-shaped," or "saddle-shaped."

Because of the small size of θ, the local radius-of-curvature r follows from Fig. 4.16a:

$$r = D/2\theta \qquad\qquad (4.20)$$

if D is the distance between the contours.

It is interesting to note that in some cases, next to the broad high-intensity main contours, narrow low-intensity weaker (subsidiary) contours can be seen. (See also Fig. 4.15) Their distance from the center of the main contour may be called x. If a uniform radius-of-curvature r is assumed, such a subsidiary contour has to originate from an orientation which is tilted by an angle β with respect to the exact Bragg position ($s = 0$). From the geometry of Fig. 4.16 it follows that $\beta = x/r$. x as well as D can be measured on the micrograph. From these, r can be determined and, as a consequence, β is also known. It turns out that this subsidiary contour corresponds exactly to the secondary maximum (Fig. 4.9b) expected from the theory; in this case, the deviation parameter s is varied simply by the local foil distortion. If the kinematical conditions were strictly met, the first secondary maximum, according to Fig. 4.9b, would be located at $s = 1.5/a$ (nm^{-1}). This s equals the product βg [analogous to Eq. (4.11), Fig. 4.8]. Thus the foil thickness a could be determined. However, the prerequisites of the kinematical theory are usually not satisfied (except in extremely thin foil edges). This is also indicated by the fact that the subsidiary contours are considerably stronger than the 4.5% expected from Eq. (4.14) for the first secondary maximum in Fig. 4.9b. Qualitatively, however, the secondary maxima are also found when the problem is treated with the dynamical theory. In that case, the secondary maxima are stronger, as is borne out by Fig. 4.15. Section 4.5.5 will give an analogous discussion concerning layer thickness determination.

4.5.5. Thickness Determination Based on Section 4.5.4

a) Wedge fringes

According to Sections 4.4 and 4.5.3, the foil thickness can be determined by counting the bright or the dark fringes starting from the foil edge. From one extreme value to the next, the foil thickness increases by the *extinction distance* ξ_g^{eff} (which defines the depth periodicity). The extinction distance is given by Eq. (4.19). Obviously, before ξ_g can be read from Table 4.2, the reflection **g** producing the fringes has to be determined from the diffraction pattern; for example, in Fig. 4.13, $g = (111)$. The value for s needed for Eq. (4.19) can also be determined experimentally (according to Section 4.5.1), if Kikuchi lines are present. In practical cases, however, wedge fringes of such high intensity, as those in Figs. 4.13 and 4.14, occur only if dynamical conditions ($s = 0$) are present for the sample area. This is documented in Fig. 4.13 by the diffraction pattern which shows the (111) two-beam case. In this case, therefore, $\xi_g^{\mathrm{eff}} = \xi_g$, and the foil thickness increases from one fringe to the next (see Table 4.2) by about 55 nm (ξ_g for aluminum and $g = 111$). At the lower image border, therefore, the foil thickness is approximately $6 \times 55 = 330$ nm. Such thickness of Al is known from experience to be close to the limit of electron transparency. It is difficult to obtain electron transmission with 100 kV electrons through thicker foils. The calculation thus shows that the foil in Fig. 4.13 has a "steep" edge, since at a 2 μm distance from the edge it is not transparent any more.

b) Secondary maxima of bend contours

In Section 4.5.4 it has been shown that at the locus of the first secondary maximum the deviation parameter s (in reciprocal space) is linked to the distance x between main and secondary maximum (in real space). The linkage is expressed by the following equation:

$$s = \beta g = x\, g/r,$$

where r is the radius of curvature and $|\mathbf{g}| = g$ the reciprocal lattice vector. If Eq. (4.20) is substituted for r, one obtains $s = 2\theta xg/D$. With $g = 1/d$ and $\lambda = 2d \sin \theta \approx 2d\theta$, one obtains the equation for calculating s:

$$s_{\text{(first secondary maximum)}} = x\lambda/Dd^2 \; [\text{nm}^{-1}], \qquad (4.21)$$

where d is the interplanar spacing for that reflection causing the bend contour, λ is the wavelength, and x, as well as D, is defined by Fig. 4.16. As already mentioned in Section 4.5.4, with the experimentally determined s at the locus of the first secondary maximum, the foil thickness would be $a = 1.5/s$ according to Fig. 4.9b, if strictly kinematical conditions would prevail. A more exact treatment, however, yields the following formula, valid for both the dynamical and the kinematical range (after Hirsch *et al.* [1, p. 417]). The 1st, 2nd, 3rd, . . . nth secondary maxima appear when s assumes the values given by the equation

$$a = \frac{n}{(s^2 + 1/\xi_g^2)^{1/2}} . \qquad (4.22)$$

With $n = 1$ (for the first secondary maximum) and the value of s present at that maximum from (4.21), the foil thickness a is obtained [6].

For constant foil thickness, the distance of the secondary maxima from the main maximum, i.e., x, *increases with increasing radius-of-curvature* r [for each secondary maximum, $x = \beta r$ (with $\beta = s/g$) is constant for any given thickness a, see Eq. (4.14)]. On the other hand, for a *constant radius-of-curvature* r, *the distance of the secondary maximum from the main maximum increases with decreasing foil thickness* a [according to Eq. (4.22)], the larger s, the smaller a, and, because of $s = xg/r$, the distance x increases with increasing s, if r is constant. The distance D, however, between the symmetrical main maxima (bend contours) in Fig. 4.16a depends only on the radius-of-curvature r and not on the thickness a [Eq. (4.20)].

In summary it should be noted that the methods for determining the foil thickness, discussed in this chapter, are new methods, i.e., they are, in principle, independent of the trace method (Section 3.9). They have a definitely inferior accuracy, compared with the trace method. The reason for this is the uncertainty concerning the exact values of ξ_g and the low measuring accuracy for the distance x. In addition, in many cases the radius of curvature is not uniform. Therefore, the thickness value a should be viewed only as an approximation, while with the trace method accuracies of 5% and better can be attained. However, the traces necessary for this method are not always present in a sample.

4.6. Application of the Basic Equation to Real Crystals

Real crystals differ from ideal crystals in that a significant proportion of the lattice atoms in the former are *not* located at the ideal lattice site r_n, defined in Eq. (3.3). Instead, because of lattice imperfections, they are displaced from their normal position by a displacement vector v. Therefore, if the intensity I_r, scattered by real crystals, or the amplitude A are to be calculated, the vector r for each atom in Eq. (4.5) has to be replaced with $r + v$. The basic equation [Eq. (4.5) together with Eq. (4.9)] thus is modified:

$$A (s) = f(\theta) \int_{\text{crystal}} \exp[2\pi i(g + s)(r + v)] \, dV. \qquad (4.23)$$

The product $g \cdot r$ is an integer (see Section 4.3), consequently $\exp(2\pi i \, g \, r) = 1$. Since $s \ll g$[6] and $v \ll r$, the product $s \cdot v$ is negligible and one obtains

$$A (s) = f(\theta) \int_{\text{crystal}} \exp[2\pi i g v] \exp[2\pi i s r] \, dV. \qquad (4.24)$$

This equation, as is Eq. (4.5), is valid for the scattered intensity of an arbitrary crystal volume K which is irradiated completely by an electron beam. In Section 4.5.2 it was found, however, that for a *thin foil* the reciprocal lattice regions capable of reflection are merely *streaks* in a direction parallel to the primary beam k_0, i.e., A equals zero everywhere except in those directions k which represent an intersection with the Ewald sphere (Fig. 4.10).

From now on, to conform with the customary usage in the literature, the coordinate system of Fig. 4.10 is positioned in such a way that the x–y plane is parallel to the plane of the foil and the z direction is parallel

---

[6] g has the order of magnitude 10 nm^{-1}, s in the example of Section 4.5.1 is a few $1/100$ nm^{-1}.

to the primary beam $\mathbf{k}_0$. Thus, for the case of a thin foil, $\mathbf{s}$ has only the component s_z, i.e., $|\mathbf{s}| = s = s_z$. The equation for a thin foil with lattice imperfections then becomes

$$A\,(s) = f(\theta) \int\!\!\!\int\!\!\!\int_K \exp[i\phi(x,y,z)]\,\exp[2\pi i s z]\,dx\,dy\,dz \qquad (4.25)$$

$$\boxed{\text{diffraction intensity}}$$

The abbreviation

$$\boxed{\phi\,(x,y,z) = 2\pi \mathbf{g}\,\mathbf{v}\,(x,y,z)} \qquad (4.26)$$

is the *phase shift* in the scattered wave caused by the shift $\mathbf{v}(x,y,z)$ of the atoms away from their normal positions.

The integration volume K in Eq. (4.25) is that cylinder whose diameter is given by the beam diameter in the sample (e.g., 1 or 2 μm, depending on the selected area aperture used), and whose height is the foil thickness (e.g., 0.1 μm). So far, the purpose of calculating the amplitude A or intensity I_r has been to determine the *total* intensity scattered by the volume K into a diffraction spot $g(hkl)$, i.e., the *intensity of just that diffraction spot*. On the other hand, to determine the *local image contrast* at a given place (x,y), a different type of calculation is necessary. According to Section 1.6, the intensity I_r at the lower foil surface has to be calculated for *only this place* (x,y), since the image contrast *at every place* is determined by the transmitted electrons exiting from the lower sample surface and passing through the objective aperture. These electrons have the intensity $I_d = I_0 - I_r$. To calculate this intensity, one imagines the total cylinder K to be subdivided into many parallel *columns*. If Eq. (4.25) is applied to each of these columns at a point (x,y), one obtains, as desired, $I_r(x,y)$ or $A(x,y)$. For this so-called *column approximation* the assumption is made that each column can be integrated separately, i.e., neighboring columns are independent of each other, and that the conditions within each surface element $(x + dx, y + dy)$ of the column (column cross section) are constant. To guarantee the latter stipulation, one imagines the column cross section to have the order of magnitude of the atomic distances or the lattice parameter.

Under these conditions, only dz remains of the former integration variable dV and one obtains for $A(x,y,s)$:

$$A\,(x,y,s) = f(\theta) \int_{\text{column}} \exp[i\phi(x,y,z)]\,\exp[2\pi i s z]\,dz \qquad (4.27)$$

$$\boxed{\text{local image contrast}}$$

4.6.1. Dislocations

4.6.1.1. Derivation of the contrast of a screw dislocation

The *screw dislocation* is a simple, yet important, example for the kinematical contrast theory of imperfect crystals. Assume a screw dislocation line extends from G to H parallel to the foil surface in the crystal foil under examination (Fig. 4.17). The dislocation core is designated with σ. The coordinate system x,y,z, as drawn in Fig. 17, is placed with its origin in the dislocation line, so that the Burgers vector $\mathbf{b}$ is antiparallel to y, and the positive z axis points downward (parallel to the primary beam). The dislocation line does not have to lie in the center of the foil; the latter can extend from $-z_1$ to $+z_2$.

Atoms originally located on a circle are forced by the screw dislocation into a helix which is drawn around G in the figure. As a consequence, the atoms originally located within a *column* parallel to the primary beam are now located in the *distorted* column designated by A–B–C. Assume the distance between σ and the column is x. As can be seen, during a circuit around σ, the displacement vector $\mathbf{v}$ increases continuously from zero (for $z = 0$, $\alpha = 0$) to the value of the Burgers vector $\mathbf{b}$, i.e.,

$$\mathbf{v} - \mathbf{b}\,\frac{\alpha}{2\pi} = \frac{\mathbf{b}}{2\pi}\arctan\frac{z}{x}, \qquad \text{for } 0 \leqslant \alpha \leqslant 2\pi.$$

Consequently Eq. (4.26) changes to

$$\phi\,(x,y,z) = \mathbf{g}\cdot\mathbf{b}\,\arctan\,(z/x) = n\,\arctan\,(z/x), \qquad \text{with } n = \text{integer},$$

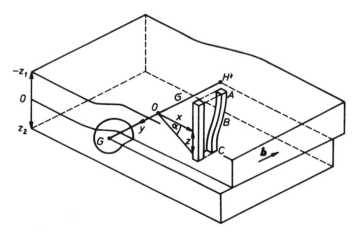

Fig. 4.17. Screw dislocation G–H in a crystal foil. For details see text (modified after Hirsch *et al.* [7]).

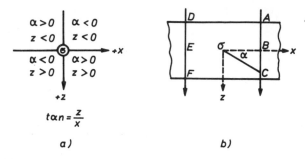

Fig. 4.18. a) Sign of the integrand in Eq. (4.28) around a screw dislocation (after Alexander [3]). b) Screw dislocation σ in a crystal foil (cross section).

if **b** is a lattice vector. For complete dislocations, this is always the case. Thus, for the usual reflections **g** with fairly low indices, $n = \mathbf{g} \cdot \mathbf{b}$ can, for example, have the values 0, ± 1, ± 2, or ± 3. (In the fcc lattice, $\mathbf{b} = \frac{1}{2}\mathbf{a}\langle 110\rangle$.) From Eq. (4.27) one obtains the local scattering amplitude

$$A(x,y,s) = f(\theta) \int_{-z_1}^{z_2} \exp[i\, n \arctan (z/x)] \exp[2\pi\, isz]\, dz. \qquad (4.28)$$

As expected, there is no dependence on y; however, for the following, one should keep in mind the dependence on n (depending on the particular reflection **g** used for observation).

The integrand in Eq. (4.28) can be visualized in the following way: Without a dislocation ($b = 0$, $n = 0$), only $\exp[2\pi isz]$ remains. If also $s = 0$, we have the simplest case where the scattering amplitudes of the N atoms stacked in the column are added, resulting in $A = N f(\theta)$. If s is finite, each subsequent atom in the column causes a certain *phase shift* φ of the wave [compare Eq. (4.1)!] which is defined by the expression $\exp(2\pi isz)$. The *amplitude–phase diagram* (APD) of the perfect crystal is, as derived in Section 4.4, a *circle* whose radius r, according to Eq. (4.16), has the value $r = 1/(2\pi s)$.

If a dislocation is present, the first exponential factor in Eq. (4.28) causes for each "elevation" z in the column an additional phase shift of the secondary wave. This shift may *increase or decrease the local radius of curvature of the* APD *circle*. The effect can easily be interpreted[7] with the help of Fig. 4.18. Figure 4.18a shows the signs of $\alpha = \arctan(z/x)$ and of z in the four quadrants of the space around the dislocation. In two adjacent quadrants ($x > 0$), α has the same sign as z. Here the phase difference caused by the dislocation (first factor in the integrand, α) in-

---

[7] After Alexander [3].

creases the phase difference between secondary waves, present for the perfect crystal (second factor in the integrand, z), if n and s have the same sign. On the other side of the dislocation ($x < 0$, column D–E–F in Fig. 4.18b), the effect is reversed.

The amplitude–phase diagrams in Fig. 4.19 represent a beam to the right and left, respectively, of the dislocation as designated in Fig. 4.18b. For the subsequent discussion, we assume $n = 1$ and $s > 0$. At a large distance from the dislocation, the APD are circles, drawn as dashed lines. If the beam moves closer to the dislocation, and for $x > 0$, the radius of curvature *decreases* because the phase shift increases. This produces the more *"tightly coiled"* spiral (compared to the circle) of Fig. 4.19b along the path ABC. Conversely, for $x < 0$, the circle expands into the *larger* spiral DEF, see Fig. 4.19a. Since the resultant amplitude, after passing through the column, is determined by the *cord* D–F or A–C, respectively, in the APD, the essential result is: For the case $x > 0$, the intensity of reflection is not greater than that from a perfect crystal; for $x < 0$, on the other hand, reflection is stronger than in a crystal without dislocation. *In bright field, the dislocation is imaged as a dark line which accompanies the dislocation, but is located asymmetrically on one side of its core.*

The nonhorizontal (not symmetrical) position of the cords D–F and A–C in Fig. 4.19 is to signify that the dislocation is not necessarily in the center of the foil.

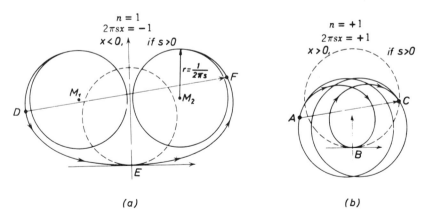

(a)   (b)

Fig. 4.19. APD for a crystal column near a screw dislocation (Fig. 4.17). The dashed circle applies to an ideal crystal. The amplitude scattered by the column ABC in Fig. 4.17 is given by the *cord connecting* the points A, C, or D, F (for $x < 0$). A, C and D, F correspond to upper and lower surface of the crystal, respectively. The distance $A - C$ or $D - F$ *measured along the curves* equals the crystal thickness. The amplitude scattered from one side of the dislocation a) is greater than that scattered from the other side b). For each value of $2\pi sx$, analogous APD can be constructed (after Hirsch *et al.* [7]).

The principal effect of the dislocation is the mutual displacement of the centers M_1 and M_2 of the original and the final circle in Fig. 4.19. This displacement depends on $2\pi s x$. The mean amplitude (i.e., disregarding the depth oscillations) is determined by the distance $M_1 - M_2$. The corresponding intensities have been calculated [7] as a function of $2\pi s x$ and are plotted in Fig. 4.20. This figure also shows quantitatively the asymmetry of the image contrast of a screw dislocation, corroborating the qualitative derivation by means of the APD.

The curves in Fig. 4.20 are proportional to the intensity. For constant s, the curves furnish a measure of the location dependence x of the intensity. To obtain the latter, the curves have to be multiplied with $(2\xi_g s)^{-2}$ [1]. However, ξ_g increases for increasing values of $n = \mathbf{g} \cdot \mathbf{b}$, i.e., for higher indexed reflections $\mathbf{g}$. Consequently, the relative heights of the

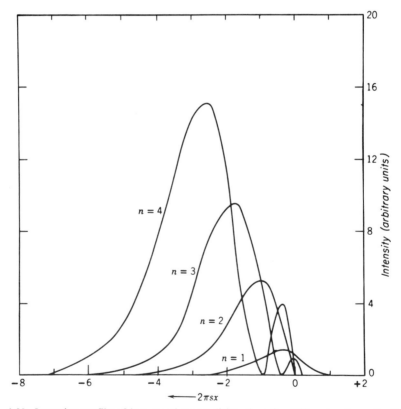

Fig. 4.20. Intensity profile of images of screw dislocations for different values of n (i.e., $\mathbf{gb}$). The dislocation core is at $2\pi s x = 0$. The contrast is always asymmetrical with respect to the dislocation core (after Hirsch *et al.* [7]).

curves are not realistic for increasing n. For reliable estimates of the intensities, the dynamical theory should be used [1].

If, on the other hand, the curves in Fig. 4.20 are used with constant x, one sees that the dislocation contrast is largest for a small tilt s away from the exact Bragg position of a reflection **g**; with further tilt the contrast drops again to zero. For very large deviations s, the maximum would move very near to zero, and the intensity would be so small that the dislocation is not resolvable. In addition, according to Fig. 4.9b, the scattering intensity of the perfect crystal decreases rapidly with large s. For $s = 0$, on the other hand, the kinematical theory is not valid, as discussed in Section 4.5; one has to use dynamical calculations (e.g., Hirsch *et al.* [1]). For "intermediate" deviations s, e.g., $s = 0.03$ nm^{-1} (compare Section 4.5.1), one can determine from the curves that the contrast maximum, at about $|2\pi s x| \approx 2$, is located to one side of the dislocation at a distance $x \approx 10$ nm. The intensity profile, according to Fig. 4.20, has about the same *width*. Both values agree with experimental experience. One sees from these results that it is important to tilt the sample in order to obtain optimum contrast.

4.6.1.2. The g·b criterion

The scalar product **g·b** $= n$ in Eq. (4.28) equals 0 not only when **b** $= 0$, but also when **b** is normal to **g**. In that case, the amplitude of the ideal crystal is all that remains in Eq. (4.28), i.e., the *dislocation is invisible if it is imaged with a reflection* **g** *for which* **g·b** $= 0$. (See Fig. 4.25, to be discussed later.) This fact can easily be visualized; it means: *a screw dislocation is invisible if the Burgers vector* **b** *is parallel to the reflecting lattice plane*. In the case shown in Fig. 4.17, this would apply to all lattice planes which contain the dislocation line as a zone axis. These planes are not "bent" as a result of the dislocation, in contrast to the (x,z) plane. For the latter, **g**‖**b**, and therefore n and the contrast are maxima.

To generalize this fact: The contrast is *not* affected for all those lattice defects whose displacement vectors **v** are exclusively parallel to the reflecting lattice plane. On the other hand, the contrast is a maximum for **v**‖**g**. Or, in other words: for electron diffraction, it matters only whether or not the diffracting lattice planes are perfectly plane; it does not matter whether atoms are displaced parallel to the diffracting lattice planes. In this sense, the asymmetrical maxima of Fig. 4.20 can also be interpreted as follows: When s or x are varied, the curvature of the column *ABC* (local lattice bending) causes the reflection conditions to be either better or worse than are those for the undisturbed lattice. This will become even clearer for the case of edge dislocations (Section 4.6.1.3).

The fact that dislocation contrast can disappear if $\mathbf{g} \cdot \mathbf{b} = 0$, i.e., that the dislocation is invisible,[8] has two important consequences:

a) It is the basis for procedures for the determination of the Burgers vector $\mathbf{b}$ (see Section 4.6.1.5).

b) It congently shows how necessary it is to produce different orientations s by tilting the sample during practical microscopy. Only then can one be sure not to overlook dislocations or other details for lack of contrast and to obtain optimum contrast. According to Section 4.6.1.1, the latter is present especially for small values of s, i.e., at the edges of extinction contours.

4.6.1.3. Edge dislocations

For edge dislocations, the construction of amplitude–phase diagrams and the calculation of intensities on the basis of Eq. (4.27) can be treated in analogy to the corresponding procedures for screw dislocations. Therefore, details are omitted here. The more important results are

a) One obtains a similar *asymmetry* (lateral displacement) of the dislocation image compared with the true location of the dislocation line. The asymmetry as well as the width of the dislocation line image are approximately 5–15 nm.

b) The effect of the $\mathbf{g} \cdot \mathbf{b}$ criterion for different dislocation positions within the foil is illustrated in Fig. 4.21. If $\mathbf{g} \cdot \mathbf{b} = 0$, the dislocation image does not disappear completely, but a very small contrast is still present. While for screw dislocations all vectors $\mathbf{v}$ are entirely parallel to $\mathbf{b}$, for edge dislocations small components of $\mathbf{v}$ lie in directions deviating from $\mathbf{b}$ (slight bending of lattice planes *normal to* the inserted half-plane).

c) The effect of an edge dislocation can qualitatively be visualized with the help of Fig. 4.22. Assume that the lattice planes of the undisturbed crystal are exactly parallel to the primary beam $\mathbf{k}_0$. According to Section 3.11, this is the *symmetrical Laue case,* and, according to Section 4.5.1, s in this case is always negative. The following discussion treats the conditions in a dark field image taken with the reflection $\mathbf{g}$. The exact *Bragg* position of those lattice planes which *would* correspond to this reflection $\mathbf{g}$, is drawn in Fig. 4.22, although the *sample,* according to the above assumption, is positioned for the symmetrical *Laue* case. As discussed in Section 3.7, the effective $\mathbf{g}$ vector can be determined from the diffraction pattern, after proper correction for magnetic rotation. Thus the signs of $\mathbf{s} < 0$ and $\mathbf{g}$ are known.

---

[8] Prerequisite is the two-beam case; for the multibeam case, of course, the condition $\mathbf{g} \cdot \mathbf{b} = 0$ cannot be fulfilled for all $\mathbf{g}$.

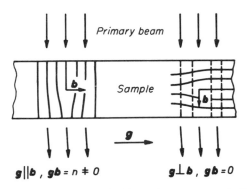

Fig. 4.21. *Left:* Edge dislocation is imaged with maximum contrast (reflecting lattice planes are severely distorted). *Right:* Edge dislocation is nearly without contrast (reflecting lattice planes, dashed, are very little distorted).

The drawing shows how the dislocation image is shifted with respect to the true location of the dislocation. Take the dislocation at the left in the figure. The first few lattice planes to the *right* of the dislocation core are bent in such a way that locally their position approximates (is parallel to) the ideal Bragg position. This match is better than that for the undisturbed crystal and even more so if compared with the lattice planes to the left of the dislocation. Therefore the highest intensity with the angle 2θ (bright in the dark field, dark in the bright field image) will be produced slightly to the right of the dislocation core. For the dislocation drawn in the right half of the figure, the reverse is true if the same reflection **g** is

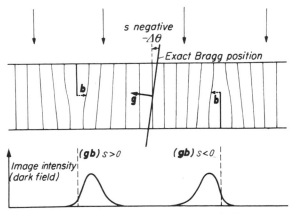

Fig. 4.22. Sketch to help visualize the relationships between the Burgers vector **b**, the normal to the reflecting lattice plane **g**, the sign of the tilt s, and the position of the dislocation image with respect to the true locus of the dislocation (after Reimer [8]).

used. Thus, the asymmetry derived (in the last section) analytically and graphically for the screw dislocation can be visualized more simply for the edge dislocation.

Since for the left dislocation, **g** and **b** have opposite signs, the product (**g·b**) s > 0; the reverse is true for the right dislocation. In the example discussed, the signs of **s** and **g** are known or have been determined from the diffraction pattern. Consequently, the *sign of b* can be deduced from the asymmetry, i.e., one can determine whether the extra half plane is in the upper or lower half of the foil (with respect to the primary beam **k₀**).

Determination of the sign of s from interference fringes

If the sample is *not* in the symmetrical Laue position, *s* can be calculated with Eq. (4.21) if a *pair* of bend contours is present. In this case, *x* is the distance between dislocation and center of the contour. The correct sign of *s* follows from Fig. 4.16. If only *one* of the contours of Fig. 4.16 is present, the sign of *s* can be determined with the following experiment [8]: According to the convention for the sign of *s* (Section 4.5.1), $s > 0$ if the rel-point **g** lies inside the Ewald sphere, i.e., angle of incidence > Bragg angle θ. With the help of the dark field image, one tries to visualize the position of the reflecting lattice planes. Then one *tilts* the sample in such a direction that the angle of incidence increases. If *s is positive* during this maneuver, the interference fringe *moves away* from the (stationary) dislocation, and vice versa. The proof for this assertion can be taken from Fig. 4.16.

If no interference fringes are present, it is possible to determine *s* from Kikuchi lines (Section 4.5.1).

The determination of the sign of **b** discussed above is a *special case of Burgers vector determination* (Section 4.6.1.5).

It has already been emphasized that image contrast from dislocations (as well as from other lattice defects or small object details) is particularly strong for small values of *s*, i.e., near extinction contours. This is due to the fact that contrast and intensity vary with s^{-2}, according to Eq. (4.14), Fig. 4.9b. According to Fig. 4.16 and Section 4.5.4, *the parameter s reverses its sign when it crosses an extinction contour*. Thus, it would be expected that the image of a dislocation lying asymmetrically on one side would change over to the other side when the dislocation crosses a contour (Fig. 4.23). This effect has been observed experimentally (Fig. 4.24); it is the simplest and often the only way to demonstrate experimentally the frequently mentioned asymmetry. If the sign of the asymmetry can be recognized in this manner, then, as was proved above, the sign of **b** can also be determined.

For $s = 0$, where the reflection is strongest, i.e., in the center of an extinction contour, the conditions of the dynamical theory prevail. Without going into the details of proof, it should be mentioned here that in the dynamical case the image of the dislocation for n = 1

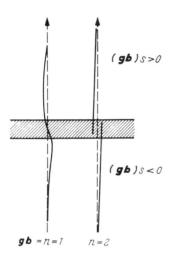

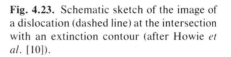

Fig. 4.23. Schematic sketch of the image of a dislocation (dashed line) at the intersection with an extinction contour (after Howie *et al.* [10]).

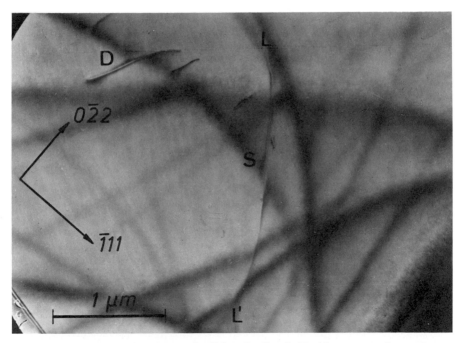

Fig. 4.24. Asymmetry of the image of a dislocation line $L–L'$ with respect to the true locus of the dislocation core. On intersecting an extinction contour S (change of sign of s) the asymmetry, too, changes its sign. Foil orientation (211). *Double contrast* of a dislocation at D because of multiple beam condition. *Material:* Al 99.99, annealed and slightly elongated, $30,000 \times$.

is located at the "correct place" at $x = 0$. A prerequisite is that the dislocation is in the middle of the foil. For $n = 2$ and $s = 0$, the dislocation image is a symmetrical double line (for $n = 3$ a triple line, etc.), as is indicated in Fig. 4.23. Thus one can distinguish between $n = 1$ and $n = 2$, if the dislocation image in the center of the contour is sufficiently resolvable.

4.6.1.4. Double contrast and more complicated cases of dislocation contrast

The asymmetry of the dislocation image has been treated above in detail for the cases of pure screw or pure edge dislocations. Similar results were obtained for dislocations which have a mixed character with respect to **b**. All derivations were valid only for the two-beam case, where imaging takes place in the light of only *one* reflection **g** which enters into the equations. If in practical cases additional reflections occur with intensities negligible compared to that of the main reflection, those extra reflections also play a negligible role in image contrast. If, however, a genuine multibeam case is present, the **g** vectors lying in different directions can produce in bright field an image on each side of the dislocation line. The (trivial) *double contrast* created in this manner can easily be identified by taking a dark field image with one of the individual reflections. If the line remains double, two possibilities exist: Either two closely spaced parallel dislocations are actually present (so-called *pair dislocations* which, e.g., occur in ordered alloys), or the lines are the two contrast maxima of *one* dislocation, such as would be expected theoretically for $n = 3$ or $n = 4$ (Fig. 4.20). In the latter case, the two maxima have very different intensity and thus can be distinguished from those of pair dislocations.

Finally, it should be pointed out that in the limited space of this volume only the most important cases of dislocation contrast could be treated with examples. Many more cases have been calculated and observed, e.g., dislocations of arbitrary character traversing the foil obliquely or vertically. In the dynamical theory, the contrast depends sensitively on the distance of the dislocation from the surface. Also, dark field and bright field images are no longer complementary to one another. For dislocations traversing a foil obliquely, periodic changes of contrast can occur which make the dislocations appear with *interrupted, zigzag, or alternating light–dark contrast* [10]. Screw dislocations intersecting the foil surface at right angles will produce a point-shaped contrast despite $\mathbf{g} \cdot \mathbf{b} = 0$.

Dislocation loops [10] are special cases. Because of the asymmetry, the ring diameter of the loop image can be larger or smaller than the true diameter of the loop. The change again occurs when either **s** changes (at a contour) or **g** changes (when the opposite reflection is chosen for the dark field image). The reader may want to consider as an exercise how he can determine, in this manner, whether the dislocation loop is a vacancy loop or an interstitial loop [analogous to the determination of the sign of **b** for an edge dislocation from (**g·b**)s, Section 4.6.1.3].

For further details consult the extensive book literature [1] where many original papers are also referenced.

A newer method should be mentioned here: Under given conditions, electron microscopic contrasts can be calculated by computer, using the kinematical as well as the dynamical theory (computer-simulated contrast prediction). Specifically for different types of dislocations, such investigations [25–27] show good agreement between calculated and observed contrasts.

Finally, the so-called *"weak-beam method"* has brought progress concerning *better resolution* of dislocation contrast [28, 29, 32]. With this method, dark field images are made not with one of the strongest diffraction spots but with a higher-indexed weaker reflection. It can be shown theoretically that sharper dark field images can thus be obtained. In addition, because the parameter s is larger for such images (see Section 4.5.1 and Table 4.1), frequently kinematical calculations are still applicable, thus simplifying the analyses.

4.6.1.5. Determination of Burgers vectors b

a) Direction of b

By suitable adjustment of the sample orientation (best done with a double tilting stage), in many cases one can make the dislocation contrast disappear. If the two-beam condition prevails or if the observation is made in dark field, one can determine, from **g·b** = 0, *the plane* which is normal to **g** and which must contain the Burgers vector **b**, but one cannot determine **b** itself. To determine a completely unknown Burgers vector, one has to try to find (by tilt adjustment) *two* different orientations so that for each of the imaging reflections $\mathbf{g}_1$ and $\mathbf{g}_2$ the dislocation contrast disappears. The unknown vector **b** is the vector product $\mathbf{b} = \mathbf{g}_1 \times \mathbf{g}_2$.

This most general of all procedures is very time-consuming. In practice, however, **b** usually is not completely unknown. Burgers vectors can have only a limited number of values determined by the crystal structure. Consequently, one knows for a particular experiment that, e.g., slip in fcc lattices will produce perfect dislocations with $\mathbf{b} = a/2 \langle 110 \rangle$ or, in a different example, one can expect half-dislocations bordering a stacking fault. The task then consists in determining, e.g., for the perfect dislocation, which of the six crystallographically possible $\langle 110 \rangle$ directions is the direction of the actually present **b** vector.

Perfect dislocations

Consider, e.g., a foil of a fcc crystal parallel to (111). If this was the slip plane, the first three dislocations **b** listed in Table 4.4 can occur in the foil as mobile dislocations. The other three dislocation types in this case are not mobile (prismatic dislocations) because they do not lie in the (111) plane; however, they can be present. To determine which specific **b** is under observation, at least three micrographs have to be taken, each with a different reflection **g**. In each case, one has to check whether or not the dislocation contrast disappears. The reflections operative for a (111) oriented foil are of the type {220} (see Fig. 3.7). Unfortunately, these are not suitable for the test since for some of the possible **b** values, they never yield the value **g·b** = 0. (This can easily be verified.) However, one can use those {111} reflections (see Table 4.4) which can be obtained by a 19.5° tilt of the sample. (19.5° = 90° minus the angle between two {111} planes.)

By using Table 4.4, the experimental results can thus be correlated unambiguously with the type of **b**. If the dislocation contrast disappears (**g·b** = 0), for example, in the light of **g** = ($\bar{1}$11) but not in the light of ($1\bar{1}$1) and ($11\bar{1}$), the Burgers vector in question is **b** = $\frac{1}{2}[0\bar{1}1]$, etc.

As was mentioned in Section 4.6.1.3, the contrast of edge and mixed dislocations (as opposed to screw dislocations) does not completely disappear when the condition **g·b** = 0 prevails. A minimum contrast remains. This, however, is so small compared to the contrast for **g·b** = 1, that for all practical purposes the dislocation is invisible and the criterion can be used as described.

Table 4.4 Values of **g·b** for determination of Burgers vector in fcc lattices[a]

b \\ g	$1\bar{1}1$	$\bar{1}11$	$11\bar{1}$
1/2[$1\bar{1}0$]	1	−1	0
1/2[$10\bar{1}$]	0	−1	1
1/2[$0\bar{1}1$]	1	0	−1
1/2[110]	0	0	1
1/2[101]	1	0	0
1/2[011]	0	1	0

[a] After Hirsch *et al.* [1].

After the vector **b** has been determined, one can decide from its direction relative to that of the dislocation line whether the latter has *edge, screw, or mixed character*.

Figure 4.25 shows an application. The dislocation dipoles are visible with the reflection **g** = (20$\bar{2}$), but invisible with **g** = (1$\bar{1}$1). Because of this last fact, **b** could possibly have one of the following three values (according to Table 4.4): $\frac{1}{2}$[10$\bar{1}$], $\frac{1}{2}$[110], or $\frac{1}{2}$[011]. With the procedure described, i.e., by testing **g·b** with other {111} reflections, it was found that **b** = $\frac{1}{2}$[10$\bar{1}$]. Since the direction of this Burgers vector is normal to the dislocation lines, the dipoles are *edge dislocations*.

Partial dislocations

For partial dislocations **b** is not a lattice vector; it is smaller than that. For *Shockley partial dislocations*, for example, **b** = $a/6\langle112\rangle$; in a fcc lattice, it represents the boundary of a stacking fault, unless the latter borders on a grain boundary. The second important type of partial dislocations is the *Frank partial dislocation* with **b** = $a/3\langle111\rangle$. It results from the condensation of vacancies into plane "disks" on {111}; each disk is surrounded by a Frank dislocation. For partial dislocations, **g·b** has the following values, depending on **g**: 0, ±1/3, ±2/3, ±1, etc. It can be calculated that the contrast of partial dislocations does not disappear for **g·b** = 0. Rather, the contrast for **g·b** = 1/3 assumes a very small minimal value making the partial dislocation in that case practically invisible [9, 10]. An example for the disappearance of partial dislocations can be seen in Fig. 4.31. In all other respects the procedure for the determination of **b** is the same as that for perfect dislocations. (Make a table for a given case, analogous to Table 4.4, listing the above mentioned multiples of ±1/3 for **g·b**.)

b) Magnitude and sign of b.

The magnitude of **b** can be determined from the image of the dislocation in Bragg position if the dislocation line intersects an extinction contour [10]. According to Fig. 4.23, Section 4.6.1.3 (last paragraph), the dislocation image at that point appears *n*-fold (since *n* = **g·b**, |**b**| can be calculated). A method for the determination of the sign of **b** also has been given in Section 4.6.1.3.

4.6.1.6. Determination of dislocation densities

The dislocation lines seen in a TEM micrograph are the *projections* of the lines which in reality lie more or less obliquely in the foil. Their true

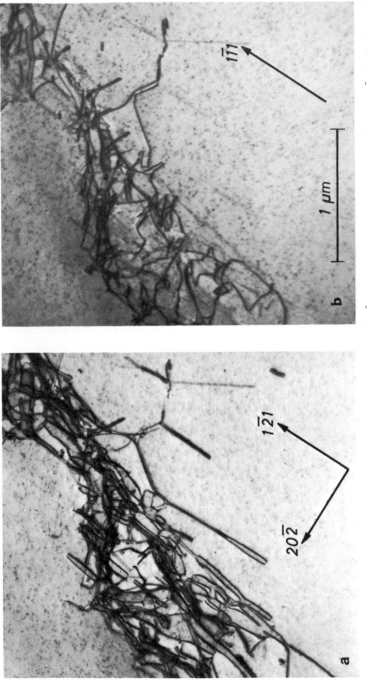

Fig. 4.25. Edge dislocation dipoles in a Cu single crystal: Burgers vector $\frac{1}{2}[10\bar{1}]$; slip plane (111). a) was imaged with $\mathbf{g} = 20\bar{2}$. the dipoles are visible. In b), imaged with $\mathbf{g} = 1\bar{1}1$, $\mathbf{g} \cdot \mathbf{b} = 0$ and the dipoles are practically invisible. In a) the primary beam is parallel to [111], for b) the sample was tilted 19.5° as explained in the text (after Hirsch *et al.* [1, p. 180], by courtesy of Her Majesty's Stationery Office).

length R can be calculated from the projected length R_p as a statistical average, if their distribution is isotropic: $R = (4/\pi)R_p$ [11]. The dislocation density ρ in cm/cm^3 is $\rho = R/(At)$, where A is the foil area used for measuring R_p and t the foil thickness. Several procedures for determining the foil thickness have already been given in Chapter 3. The measurement of the line lengths R_p, however, is a rather time-consuming process. Therefore, the following so-called line intercept method [12, 13] has been developed to expedite the determination of R_p: A number of lines of total length L is drawn onto the area A (on the micrograph). These lines can be arbitrarily or uniformly placed (they can, e.g., be straight lines or large circles). *The number N of intersections between these lines and the dislocations is counted.* R_p then is given as $R_p = \pi NA/(2L)$, and thus

$$\rho = 2N/(Lt). \tag{4.29}$$

With this procedure, dislocation densities up to 10^{11} or 10^{12} cm^{-2} can be measured.

A second method is based on the counting of the penetration points of dislocations on the foil surfaces, analogous to the etch pit method in light optics. Since in TEM the points on *both* surfaces are counted, the result has to be divided by two. For this method, the foil thickness t need not be known. However, the method can be used only for moderate dislocation densities (up to about 3×10^9cm^{-2} [14]); otherwise too many line ends (penetration points) are lost for counting because of overlap with adjacent dislocations.

Generally, however, it must be stated that *the dislocation density can be determined from TEM micrographs only with the following essential reservations:*

a) Dislocations can *be either produced or lost as a result of sample preparation*. Dislocations can be generated by plastic deformation of the delicate thin foil during handling (handling dislocations). These can generally be avoided with practice or by using the disk method of preparation (Section 2.1.4). The loss of dislocations, on the other hand, is due to the effect of *image forces* which cause dislocations near the surface to leave the foil. This effect can be estimated, e.g., by plotting the measured dislocation density as a function of foil thickness. (The use of high voltage electron microscopes, Section 1.7, is particularly advantageous for this purpose.) In some cases, the rearrangement of dislocations during thinning can be prevented by submitting the sample to neutron irradiation prior to thinning. Such "stabilization" often has been successful, e.g., for deformed Cu crystals [15].

b) Because *contrast conditions may not be fulfilled*, dislocations may be *invisible*. This can be remedied by taking several micrographs in the light of different Bragg reflections, or by taking one micrograph with several imaging reflections (multibeam case).

c) The greatest difficulty is encountered with very nonuniform dislocation distributions, when one tries to resolve dislocations in the regions of high dislocation density. This case is present, e.g., in the cell structure seen in Fig. 4.27 (compare Section 4.6.1.7). The case is similar if the dislocations occur as networks, e.g., in low-angle grain boundaries, as in Fig. 4.28. The dislocations in networks lying normal to the foil surface cannot be counted.

Consequently, the determination of the dislocation density is possible only if the dislocation distribution is reasonably uniform (as, e.g., in Fig. 4.26a) or if not too many dislocations are superimposed anywhere.

4.6.1.7. Examples of dislocation configurations

Dislocations can be arranged in very different configurations, depending on material and thermomechanical pretreatment. A very *uniform dislocation distribution* is shown, e.g., in Fig. 4.26a. It is typical of small degrees of deformation, here 3% plastic elongation, in alloys with medium and high stacking fault energy. If the stacking fault energy is low, the dislocations often lie in crowded coplanar arrays on their slip planes (e.g., Fig. 4.26b). Because they are split into half-dislocations they cannot leave the slip plane by cross slip as easily as they do in alloys with higher stacking fault energy.

Many metals with medium or high stacking fault energy will develop a so-called *cell structure* (Fig. 4.27) after medium deformations. In the cell walls the dislocation density is very high while the cell interior is nearly free of dislocations. In the cell walls, the individual dislocations usually are not resolvable. The cell structure of aluminum has been investigated [30] in more detail, especially the cell diameter and the orientation differences between neighboring cells as a function of the degree of cold deformation.

Nonuniform dislocation densities are also found in structures with *dislocation networks*, e.g., Fig. 4.28. They form very regular arrays which constitute *small-angle boundaries*. Tilt boundaries are present when a number of edge dislocations are superimposed in parallel arrangement. Two-dimensional networks of screw dislocations are twist boundaries. The smaller the width of the mesh, the larger is the angle of rotation between neighboring subgrains. Typically, dislocation networks will form during recovery after cold deformation (polygonization, subgrain structure).

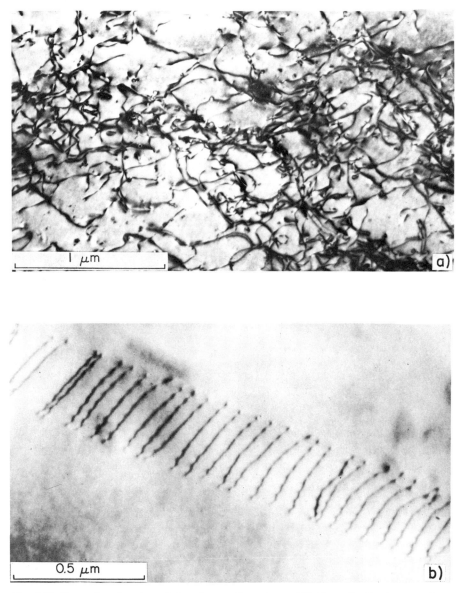

Fig. 4.26. Examples of various dislocation configurations: a) Uniform distribution over the whole field of view. *Material:* Al–5% Zn–1.5% Mg. Heat treatment: 1 hr 550°/H$_2$O + 1d RT + 2d 70°C + 3% plastic elongation, 40,000× (micrograph: R. Reichel). b) Dislocations concentrated on a single slip plane. *Material:* Permalloy 78.7% Ni, 5.0% Mo, balance Fe. Treatment: quenched from 600°C + slight deformation, 71,500× (micrograph: Pfeifer and Pfeiffer [16]).

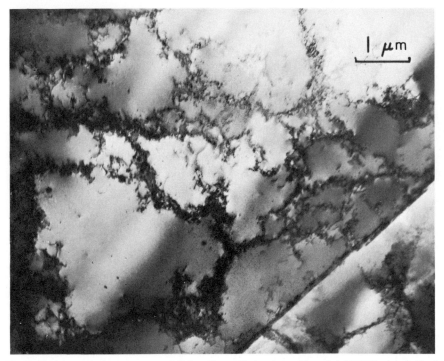

Fig. 4.27. Example for pronounced dislocation cell structure. *Material:* 99.98% nickel. Heat treatment: recrystallization anneal 1 hr 500°C + 10% plastic elongation, 15,000×.

After very *high degrees of deformation,* the dislocation density in most metals[9] increases to such an extent that the dislocations are no longer resolvable anywhere; the result is a dark cloudy microstructure. This is the case in Fig. 4.29 (area B). Such areas can easily be distinguished from already recrystallized grains (A) which are free of dislocations. The growth of such a grain into the dark deformation structure can be seen in the upper right quadrant of the micrograph. The parallel bars Z in the lower recrystallized grains are recrystallization twins typical for nickel (see also Fig. 4.14 and the remarks there concerning this material).

4.6.2. Stacking Faults

Equation (4.27) can also be applied to stacking faults. Assume that (in Fig. 4.30) a stacking fault S in a fcc crystal extends obliquely from S_1 to S_2. In each of the two halves of the crystal separated by S, crystal 1 and

[9] Exception: aluminum which exhibits the so-called "deformation polygonization."

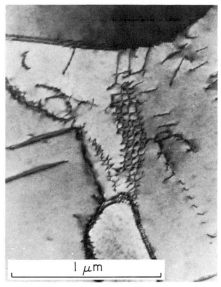

Fig. 4.28. Example for dislocation networks which form low-angle grain boundaries (subgrain structure). *Alloy:* Fe–4% Mo. Heat treatment: warm deformation by creep at 750°C (13% deformation), 40,000× (micrograph: A. Fuchs [thesis, Erlangen 1968]).

1 μm

crystal 2, the fcc lattice is perfect. The stacking fault has the effect of shifting the lattice below S with respect to the lattice above S by the amount of the Shockley partial dislocation $\mathbf{b} = a/6[112]$. This $\mathbf{b}$ is the displacement vector $\mathbf{v}$ which has to be substituted into Eq. (4.26) or (4.27) for all of the half-crystal below S. The scattering amplitude then is

$$A(x,y,s) \sim \int_{B}^{C} \exp[2\pi i s z] \, dz + \int_{C}^{D} \exp[2\pi i g b] \exp[2\pi i s z] \, dz. \quad (4.30)$$

The integration over the column B–C–D has to be done in two segments. Above S, along B–C, the lattice is undisturbed; its amplitude–phase diagram (APD) is a circle with radius $1/(2\pi s)$, Eq. (4.16). At point C the stacking fault causes a phase shift $\phi = 2\pi\mathbf{g}\cdot\mathbf{b}$. If, e.g., the reflection $\mathbf{g} = 1/a[200]$ is used, $\mathbf{g}\cdot\mathbf{b} = (1/a)[200]\cdot(a/6)[112] = \frac{1}{3}$ and $\phi = 2\pi/3$. That means, the phase shift is 120°. This angle is drawn in the APD at point C. The APD continues on a circle of the same diameter, while its tangent at C forms the angle ϕ with the first circle. The vectors correlated with the line C–D in the column are located on the arc C–D of the second circle.

The resultant amplitude $A(x,y,s)$ is the straight line $\overline{BD}$ in the APD. If the construction described is carried out for all locations x along the stacking fault from S_1 to S_2, it is found from the APD that the length of the vector $\overline{BD}$ changes periodically. The sum of the two circular arcs B–C plus C–D remains constant; only the discontinuity C shifts. Consequently,

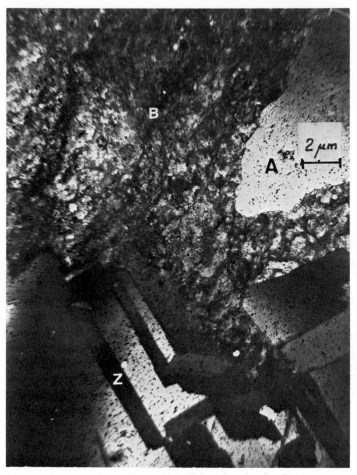

Fig. 4.29. Examples for regions of very high dislocation density B next to recrystallized grains A. Partially recrystallized condition. Bars at Z are recrystallization twins. *Material:* TD nickel (Ni–2 vol% ThO$_2$). Heat treatment: 90% cold rolled + 1 hr 700 °C (after v. Heimendahl [5]).

the stacking fault is imaged as a system of parallel bright and dark stripes. This stripe contrast at stacking faults is a special case of *wedge fringes* which have been treated in detail in Sections 4.4 and 4.5. By analysis of the APD, one obtains the stripe spacing as a function of the parameter s, similar to the derivations given in Section 4.5. (An example for stacking faults is shown in Fig. 4.31.)

Stacking fault contrast can be distinguished from the similar contrast at twin structures with the help of the diffraction pattern. Twins show

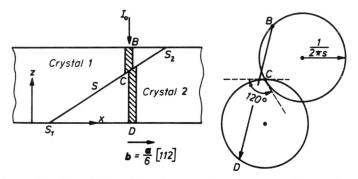

Fig. 4.30. Stacking fault S lying obliquely in a foil, and corresponding amplitude–phase diagram. Below S the fcc lattice is shifted by the amount **b**.

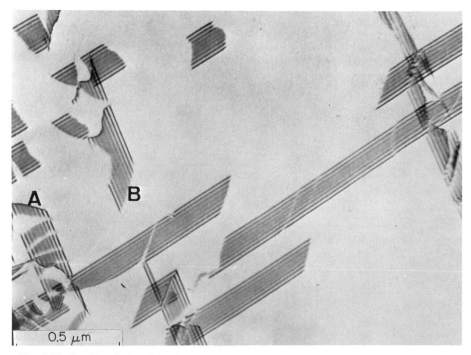

Fig. 4.31. Stacking faults with adjacent partial dislocations which are visible at A, invisible (out of contrast) at B. *Material:* Co Ni Cr Fe alloy [spring alloy], 55,000×. Heat treatment. $\frac{1}{2}$hr 1180°C/H$_2$O, annealed at 450°C (micrograph: Pfeiffer [16a]).

additional reflections (Section 4.7), stacking faults do not. Since a stacking fault is a very thin hexagonal layer, only four atom layers thick, within a fcc structure, the diffraction pattern at most will contain streaks normal to the stacking fault plane (Theorem 5, Section 4.5.2), if many such parallel hexagonal layers form packets of alternating matrix and stacking fault lamellae. This possibility to distinguish stacking faults and twins is important because frequently they will occur side by side, for example, in stainless steel or in gold leaf.

4.7. Analysis of Twin Structures

Part of a crystal is in *twin position* with respect to the matrix, if the crystal lattice is rotated by 180° around the *twin axis*. (This is not a mechanism of twin formation, it is only the description of the orientation relationship.) The *twinning plane,* normal to the twin axis, is simultaneously the habit plane of the twin crystals which frequently are plate- or lamella-shaped. With respect to the twinning plane, matrix and twin have mirror-image lattices. In a fcc crystal, any {111} plane can function as a twinning plane; in bcc crystals, the twinning plane is parallel {112}. Twins can be generated by deformation as well as by recrystallization. During deformation, the crystal yields to the shear stress by a (sudden) "flip-over" into the twin position. The direction in which this happens is called the *twinning direction.*

Twins can occur in a wide range of sizes. "Macrotwins" are visible to the naked eye or in the light microscope; in contrast, "microtwins" can be resolved only in the EM. An example for the latter is shown in Fig. 4.32. It is also crystallographically the simplest case with regard to interpretation, namely, the case where the twin axis lies in the image plane. Since the twinning plane is normal to the axis, one can recognize in the micrograph the different twins of various widths, some of them being very thin lamellae (10–250 nm). If the twins were in an oblique position, falsifying projection effects might be encountered. In the dark field image, many of the small twins can be better recognized than in bright field.

The rotation by 180° of the twin lattice also is manifested in the diffraction pattern and the reciprocal lattice. The matrix is in [01$\bar{1}$] orientation (reflections: circles). The twin reflections (crosses), or the reciprocal lattice points of the twin crystal, can be constructed by rotating the reciprocal lattice of the matrix by 180° around the twin axis. As will be recalled, the reciprocal lattice has to be imagined as being rigidly coupled with the real lattice. The 180° rotation corresponds to a reflection of all points of the matrix lattice on the twin axis. This reflection is indicated by arrows in the schematic diffraction pattern of Fig. 4.32. Besides, the

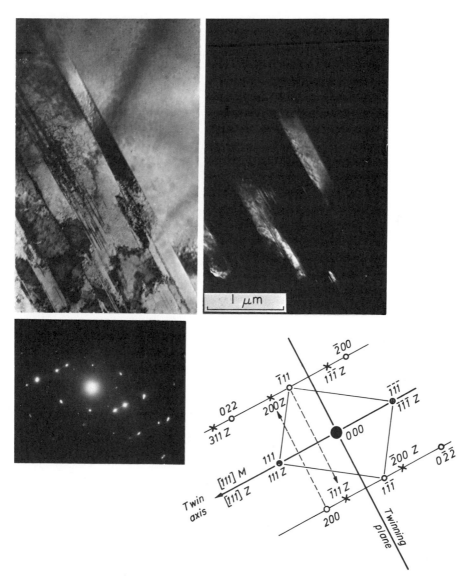

Fig. 4.32. Microtwins in stainless steel. Heat treatment: after cold rolling, annealed 10 min at 750°C. In the schematic diffraction pattern the real pattern is enlarged by the factor 2. Matrix reflections (M): circles. Twin reflections (Z): crosses. Dark field image (right) in the light of the (200) twin reflection, 20,000×.

diffraction spots of the twin lattice alone, as those of any lattice, obey the rules of Section 3.6. It can also be seen in Fig. 4.32 that the twinning plane coincides with the visible habit planes of the twin lamellae. (Allowing for the magnetic rotation, Section 3.7, the twinning plane, which is normal to the [111] axis, can be determined from the diffraction pattern.) Because some of the lamellae are very thin, the diffraction pattern shows streaks normal to the thickness direction, i.e., in [111] direction.

When the twin axis lies in the image plane, as in Fig. 4.32, the twin diffraction spots can easily be constructed with the help of the mirror reflection. The case is more complicated if the twin axis does not coincide with the image plane. Because of the 180° rotation, diffraction in the twin will occur for reciprocal lattice points which in the matrix are not positioned on the image plane (Ewald sphere). Equations can be derived which yield the position of a lattice point (hkl) after rotation by 180° around the twin axis $[pqr]$ (Section 6.3 in Hirsch et al. [1]). The twin positions thus determined are called $(h'k'l')$ and are based on the coordinate system used for indexing the matrix. Those $(h'k'l')$ which lie in the plane of the diffraction pattern identify the diffraction spots generated by the twinning process.

An example where the twinning planes lie obliquely to the image plane is given in Fig. 2.9. The matrix of the gold leaf has (001) orientation. In the diffraction pattern (lower left corner), the twin reflections are found close to the primary beam; they occur in pairs and are elongated streaks. They are close neighbors to the (010) and (01̄0) matrix reflections which are slightly more distant from the primary beam. Dark field images taken in the light of each of these individual twin reflection streaks showed either the vertical or the horizontal bars in bright contrast. In this manner, the bars have been identified as deformation twins. They can thus be distinguished from stacking faults which also occur in gold leaf. (The diffraction pattern, by the way, has been analyzed in Hirsch et al. [1], Fig. 6.19, Section 6.3.2.)

4.8. Contrast of Precipitates

The most important phenomena occurring with precipitates will be discussed with the aid of a few *aluminum alloys*, in particular the Al–4% Cu alloy. The precipitates in this alloy are similar to those in the technically significant *Duralumin type* of alloys (Al–Cu–Mg). The precipitate phases and their properties are listed in Table 4.5. The types of precipitates indicated by the sketches are representative also of those occurring in other alloys.

Table 4.5 Summary of some important types of precipitates, with Al alloys taken as examples

Type of precipitate	Fully coherent without misfit	Coherent, but with misfit on one boundary surface		Semicoherent, i.e. coherent on one interface, incoherent on the other	Incoherent
Example	GP zones in Al–Ag	GP zones in Al–Cu		θ'-phase in Al–Cu	θ-phase in Al–Cu
		GP I	GP II		
Structure	fcc, same as matrix	Cu layers, one or two atoms thick	Tetragonal $a = 0.404$ nm $c \approx 0.78$ nm	Al_2Cu tetragonal $a = 0.405$ nm $c = 0.580$ nm	Al_2Cu tetragonal $a = 0.6066$ nm $c = 0.4874$ nm
Crystallographic orientation relationship		Orthogonal axes parallel to the cube axes of the matrix			Many orientation relationships, some with high indices
Habitus	Spherical	Platelets on cube faces of the matrix			Platelets, rods, or isometrical
Size (diameter or length)	<20 nm	<10 nm	<50 nm	<1.5 µm	<8 µm
Temperature region[a]	RT–200°C	RT–160°C	100–200°C	150–400°C	150–500°C
Fig. example	—	4.33	4.33	1.13 and 3.12	4.34

[a] Valid for Al alloys with 4% Cu or 4–20% Ag, respectively.

4.8.1. Guinier–Preston Zones (GP Zones)

The early stages of decomposition often are difficult to detect even by electron microscopy. Although the *nominal resolution* of a modern EM is 0.3–0.5 nm, usually GP zones of that magnitude cannot be imaged because their *contrast* is too low (compare Section 1.5). In many cases, second phase particles ten times as large are at the *practical resolution* limit. In Fig. 4.33a, the smallest discernible GPI zones are about 0.2 mm on the micrograph, corresponding to 5 nm in the sample; however, the *individual particles* are just barely visible. Overall, they rather appear as a general *roughening* of the background.

Frequently such roughening is quoted as a criterion for the presence of GP zones. It should be emphasized, however, that this criterion should be used with caution since occasionally a certain roughening can also occur as a consequence of electrolytic polishing. Homogeneous (solution annealed) foils should be prepared for comparison; these should not show roughening. In Fig. 4.33 the zones have already attained a magnitude (5 nm) which no longer can be interpreted as "polishing effect."

In the early stages of decomposition, the *diffraction pattern* furnishes a much more reliable criterion for the presence of GPI zones. Because these are very small *platelet-shaped* particles, they cause very long streaks in the diffraction pattern. The origin of streaks has been derived in Section 4.5.2 (Theorem 4, Fig. 4.11c). Because of the extreme thinness of the GPI zones these streaks (which are inversely proportional to the thickness) extend across the whole diffraction pattern. They are easily seen in the [100] and [010] directions in Fig. 4.33a, but they would also be discernible for smaller GPI zones which cannot yet be identified with confidence in the micrograph. A subsequent solution anneal or reversion anneal will make the streaks disappear [17]. In contrast to the Al–Cu alloys, technical alloys of the Al–Cu–Mg type have rod-shaped GPB zones. Therefore, the condition coresponding to Fig. 4.33a, representing approximately the hardness maximum, cannot be detected either in the diffraction pattern or in the TEM micrograph.

Aging at somewhat higher temperatures leads to GPII zones, see Fig. 4.33b. The thickness of the platelet-shaped zones grows; consequently, the streaks become shorter. It can be calculated that in this particular case they also show characteristic interruptions. Consequently, in the reciprocal lattice, three diffuse maxima occur along a streak, approximately at positions 1/4, 1/2, and 3/4 between the matrix reflections in the $\langle 100 \rangle$ directions. These interrupted streaks are better discerned in the original diffraction pattern than on the reproduction; the schematic at the bottom of Fig. 4.33 shows their positions. They serve to distinguish reliably between GPI and GPII zones. Both kinds look the same in the TEM micrograph, except for their size.

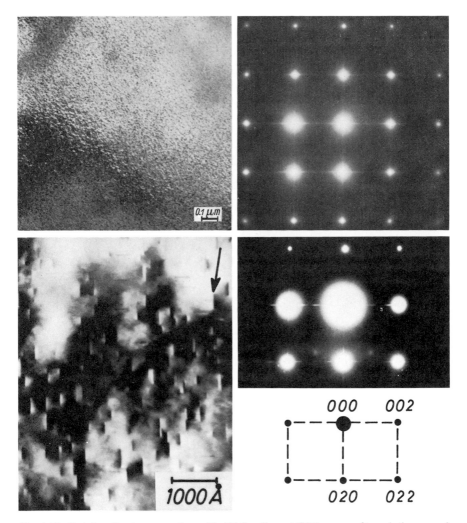

Fig. 4.33. Guinier–Preston zones in an Al–4% Cu alloy. a) GP I zones, after solution anneal + aged 2 hr at 160°C, 40,000 ×. Diffraction pattern: continuous streaks [17]. b) GP II zones, after solution anneal + aged 2.5 hr at 190°C, 120,000 ×. Diffraction pattern: interrupted streaks [18].

4.8.2. Strain Contrast

In both micrographs of Fig. 4.33, the matrix is in ⟨100⟩ orientation (cube orientation). GPII platelets occur on the three cube faces. In the image projection, two systems appear as lines at right angles to each other. The length of the lines corresponds to the platelet diameter. (The platelets of the third system, parallel to the foil surface, are so thin in the beam direction that they are unrecognizable because of insufficient contrast.) In the lower micrograph of Fig. 4.33 it is noteworthy that most of the platelets are bordered by a diffuse halo, bright on one side, dark on the other (see arrow). This is a sign of so-called *strain contrast,* typical for coherent precipitates if these have a misfit interface, see Table 4.5. The sketch in the table elucidates the *distortion of the matrix lattice planes* in the neighborhood of the precipitate. Within the region surrounded by the dashed line, the local strain field of the lattice deformation, i.e., the atom displacement **v**, is particularly large. This displacement can be treated by means of the contrast theories, just as were the lattice distortions near dislocations (compare Figs. 4.21 and 4.22). Detailed analysis, therefore, is not necessary here. With respect to lattice distortions, a platelet precipitate is analogous to a dislocation loop.

The *coherency strains* thus cause strain contrast which also is evident from the schematic in Table 4.5: On one side of the platelet, the reflection conditions for the operative diffracted beam **g** are worse than those for the undisturbed crystal, on the other side they are improved, thus causing the light–dark constrast. In Fig. 4.33b, the bright side for some GPII zones is at right, for others at the left of the zone. The reason is that this is a multibeam case (see diffraction pattern). In addition, the distribution of light and dark on both sides of the particle with strain contrast depends also on its depth location in the foil and on its size compared to the foil thickness. For these reasons, the contrast distribution may vary even for a two-beam case.

The *strain contrast* described may be so strong that it dominates the image appearance, as in Fig. 4.33b. Of course, precipitates in principle can also show the types of contrast discussed in Chapter 1, namely mass (scattering absorption) contrast and orientation (diffraction) contrast. In addition, *interface contrast* (displacement fringe contrast) may occur. This is due to the phase shift of an electron wave on passage through an interface between matrix and precipitate, provided the interface is oblique or normal to the primary beam. This effect is similar to that discussed for stacking faults (Fig. 4.30), resulting in similar fringe contrast.

4.8.3. Criteria According to Ashby and Brown [19]

For very small precipitates, the strain contrast looks like a coffee bean (⬭) showing in the center a line of no contrast, the symmetry line. This kind of contrast is present for platelets as well as for spherical particles with boundary misfit. Therefore, the two cannot be distinguished from the appearance of the image alone. According to the *first criterion of*

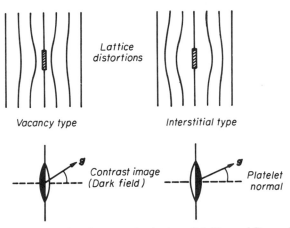

Fig. 4.34. Schematic illustrating the second criterion of Ashby and Brown [19]. From the position of **g** with respect to the light or dark half of the contrast image one can distinguish between vacancy or interstitial type of platelet precipitates.

Ashby and Brown, distinction is possible as follows: Micrographs under two-beam conditions are taken with different nonparallel reflections **g**. For spherical particles, the line of no contrast is always normal to **g**, while for platelet-shaped particles the symmetry line does not change with the direction of **g** but is fixed by the position of the platelet.

The *second criterion of Ashby and Brown* makes it possible to distinguish from the strain contrast of platelet precipitates whether these are of the "vacancy type" or the "interstitial type," see Fig. 4.34. The vacancy type is associated with "inward" lattice distortions, as seen in the sketch for Table 4.5. They occur, e.g., in GP zones in Al–Cu alloys or in vacancy condensations. Interstitial types of "outward" lattice distortions, on the other hand, are caused either by interstitial atoms or by substitutional atoms with a radius larger than that of the matrix atoms. The two types of platelet precipitate yield the same type of image similar to Fig. 4.33b and therefore cannot be distinguished without the following special technique.

According to Ashby and Brown [19], unambiguous distinction is possible *in dark field* if two-beam conditions prevail. The contrast has a bright and a dark side also in dark field. The criterion reads:

If the angle between **g** and the platelet normal is acute on the *bright* side, the platelet is of the vacancy type. If the angle between **g** and the platelet normal is acute on the *dark* side, the interstitial type is present. "Bright" and "dark" refer to the dark field image on the image screen or on a positive print.

This example shows the importance of correctly correlating **g** with the dark field image (Section 3.7) and also the usefulness of observations in a well-defined two-beam case.

4.8.4. Metastable and Stable Phases

Aging of the Al–4% Cu alloy at higher temperature will promote precipitation of the metastable θ′ phase (Table 4.5). Examples of the appearance of that phase were shown in Figs. 1.13 and 3.12. The θ′-particles no longer have strain contrast and can thus be distinguished from GPII zones [18]. The strong dark contrast of the θ′-crystals in Figs. 1.13 and 3.12 is due mainly to the higher scattering absorption in the precipitates because their Cu content (composition Al_2Cu) is higher than that of the aluminum matrix. Thus, the contrast is *mass contrast*. In addition, orientation contrast or interface contrast may also occur if the orientation is right or if the parameter *s* has the appropriate magnitude.

Differing from GP zones, the θ′-phase has a well-defined tetragonal crystal structure. Consequently, the individual precipitate particles generate their own diffraction pattern. It is a spot pattern, clearly different from the streaks typical for GP zones. Like any diffraction pattern, it can be analyzed according to the rules of Chapter 3. In this case, it can easily be distinguished from the simultaneous diffraction pattern of the matrix because the θ′-crystal has larger interplanar spacings and, consequently, a smaller diffraction pattern with the spots closer together than those of the matrix pattern. An example was shown in Fig. 1.13.

Since the θ′-platelets are precipitated only on the cube planes of the matrix, their typical regular crystallographic arrangement is very obvious in Figs. 1.13 and 3.12. In contrast, the *equilibrium phase* θ, as in Fig. 4.35, precipitating at even higher temperature, has a very irregular arrangement. Many possible orientation relationships exist between this phase and the matrix. Consequently, many habit positions are possible for the particles, which may be rod-shaped or platelet-shaped and besides may reach very different sizes. The grain boundaries and the grain boundary triple point (in the center of the micrographs) are decorated with θ particles. Nucleation of the incoherent phase is particularly facilitated there because of the severe lattice disortion in a grain boundary. In contrast to the other phases discussed, the θ crystals are so large that they can be seen in the light microscope. Comparison of the light and electron micrographs (Fig. 4.35), however, shows the considerable gain in resolution by the EM. Besides, it must be noted that the light micrograph shows an etched surface (etchant: sodium hydroxide solution, 50°C), while the EM micrograph shows a projection of the sample volume. Nevertheless, in this case the two images are very similar.

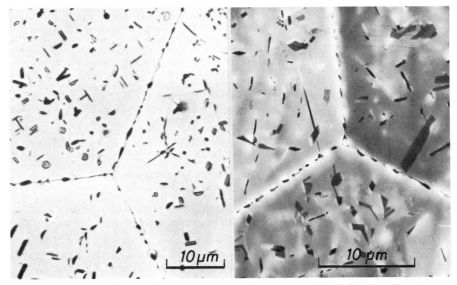

Fig. 4.35. Example of an incoherent phase: θ-precipitates in an Al–4% Cu alloy. Heat treatment: solution annealed + aged 30 min at 450°C. Comparison of light microscopic image (left) and electron microscopic image (right) demonstrates the improved resolution in the EM, even with relatively low magnification. Both micrographs are of the same microstructure but of different sample areas.

In practice, small precipitates are of considerable importance because they form obstacles for dislocation movement and thus increase the strength (age hardening). The obstruction of moving dislocations is illustrated in Fig. 4.36. At points A, dislocations have been pinned by precipitates. As a consequence, the dislocations form semicircular arcs, a process which is the basis of the *Orowan mechanism*. At points B, the dislocations have left the slip plane and have formed *prismatic dislocation loops*. At C, a slip trace is present, as in Fig. 3.13. Everywhere in the micrograph, the dislocations are very short and have been severely impeded in their movement.

Discontinuous precipitation can also be easily observed in the TEM as, e.g., in Al–Ag alloys [31].

4.8.5. Contrast Reversal at Thickness Fringes

Aluminum alloys of technical importance, besides those with Cu, are the Al–Zn–Mg alloys. Their strength is highest if they contain a mixture of GP zones and metastable hexagonal M' phase precipitates ($a = 0.496$ nm, $c = 0.868$ nm). In Fig. 4.37, the many small particles are the M' phase forming platelets on the $\{111\}$ matrix planes. Coherency strains are

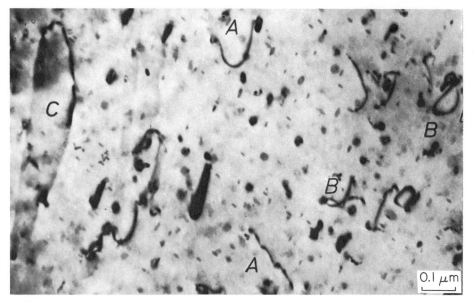

Fig. 4.36. Severe obstruction of dislocation movement by uniformly distributed nonshearable precipitates. For details see text. *Alloy:* Al–0.2% Au. Heat treatment: solution anneal + 60 hr 200°C + 5% plastic elongation, 80,000× (after v. Heimendahl [20]).

not observed. Generally (as in Fig. 4.37) the M' particles appear in *dark mass contrast* because of their Zn content. However, note that in Fig. 4.37 the particles are *bright* on one side of the thickness contours. Thus the figure is an example for the so-called *contrast reversal* at the border of thickness contours: Objects on one side of a wedge fringe have bright contrast, on the other side they have dark contrast.

This phenomenon can be interpreted only with the methods of the dynamical theory [19]. Consider the case $s = 0$ with wedge fringes present, as, e.g., in Fig. 4.13. The sample contains particles which have no strain contrast, only mass contrast. The extinction distance of the particles is designated with ξ_g^p, that of the matrix with the usual ξ_g, the foil thickness with t. If the particles have the same crystal structure as the matrix, the theory states:

1. The particles are invisible at the maxima and minima of the periodic intensity variations, i.e., for that foil thickness t, for which $t/\xi_g = 0$, 1/2, 1, . . . (compare Fig. 4.12).
2. The particles have maximum contrast on the "flanks" of the intensity oscillations, i.e., for $t/\xi_g = 1/4, 3/4, 5/4, ...$

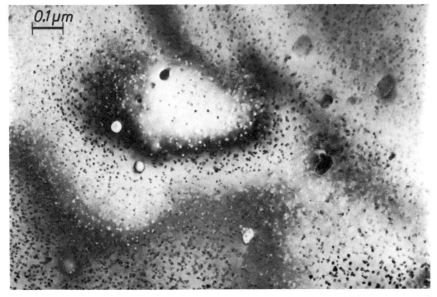

Fig. 4.37. Contrast reversal for small particles at wedge interference fringes. *Material:* Al–Zn–Mg 1. Heat treatment: 1 hr 450°C/air + 7*d* RT + 1*d* 120°C. The many small uniformly distributed precipitates (6 nm diameter) of M' phase normally appear dark because of mass contrast. However, they are bright at one side of the thickness contours. The few larger randomly distributed particles (30–80 nm) are aluminides containing foreign atoms, or insoluble inclusions (micrograph: R. Reichel and D. Puppel).

If in bright field $\xi_g > \xi_g{}^p$, then the particle contrast is

$$\left.\begin{array}{c}\text{dark}\\[4pt]\text{bright}\end{array}\right\} \text{ for } t/\xi_g = \left\{\begin{array}{l}1/4,\ 5/4,\ \ldots\\[4pt]3/4,\ 7/4,\ \ldots\end{array}\right. .$$

For $\xi_g < \xi_g{}^p$ the reverse is true.

4.9. Moiré Patterns

If two thin crystals lie one on top of the other, the superposition of the two crystal lattices usually creates complicated imaging conditions. Two special lattice relationships, however, cause interesting and easily interpreted effects: a) the two lattices lie parallel to each other and their interplanar spacings d_1 and d_2 differ only slightly, or b) the two lattices have the *same* interplanar spacing d and are rotated one against the other by a small angle ϵ. These cases are shown schematically in Fig. 4.38. One can easily produce the effect by drawing such lattices on two sheets of

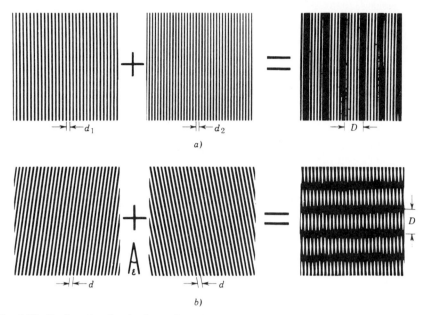

Fig. 4.38. Explanation for the formation of the two fundamental types of Moiré patterns: a) Parallel Moiré pattern with $D = d_1 d_2/(d_1 - d_2)$ and b) rotation Moiré pattern with $D = d/\epsilon$ (after Thomas *et al.* [2]).

transparent paper, placing one on top of the other, and holding them up against the light. The two lattices (as in Fig. 4.38) form a kind of "superlattice" with a much larger spacing than each of the individual lattices. Such lattices are called "Moiré lattices."

Case b) can also be demonstrated by superimposing two pieces of a fine wire mesh and slightly rotating one against the other. Case a) may be observed in daily life: If, while driving on a freeway, one approaches an overpass protected by railings with vertical bars, the two railings seem to have different "lattice constants" because of their different distance from the observer. Consequently, he sees the superlattice of Fig. 4.38, upper right; besides, it will shift during his approach.

This effect is called "Moiré pattern" because of its great similarity with the pattern of the Moiré fabric.

The lattice constant D of the Moiré lattice can easily be calculated. For the "parallel Moiré" of Fig. 4.38a, $D = d_1 d_2/|d_1 - d_2|$, and for the "rotation Moiré in Fig. 4.38b, $D = d/\epsilon$. In the latter case, the "lattice planes" of the Moiré are normal to the crystal lattice planes.

It is interesting that an edge dislocation introduced into one of the two superimposed lattices is also imaged in the Moiré pattern. Here, too, a highly magnified "dislocation" with an extra "half-plane" is seen (Fig. 4.39). As with the perfect lattices, the Moiré pattern is parallel in case a), at right angles to the original dislocation in case b).

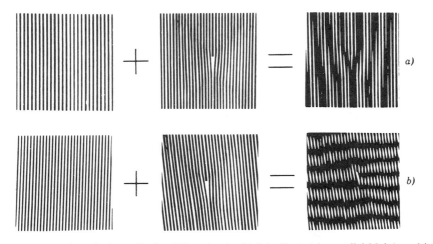

Fig. 4.39. Enlarged "image" of a dislocation by Moiré effect a) in parallel Moiré, and b) in rotation Moiré (after Thomas *et al.* [2]).

Although Figs. 4.38 and 4.39 are only optical analogies to the conditions in real atomic lattices, they nevertheless explain in first approximation the Moiré effects frequently observed in the EM when crystals are superimposed. An instructive example was Fig. 2.10 (TEM image of mica). Nearly everywhere on the micrograph, particularly at M, one can see the fine feathering of parallel lines which constitute the Moiré pattern. The mica foil consists of several superimposed lamellae of the same lattice constant, rotated against each other.

A case of parallel Moiré is shown in Fig. 4.40. Thin platelets of the metastable η'-phase have precipitated in an Al–0.2% Au alloy [20]. On their basal plane the precipitates are nearly, but not exactly, coherent with the Al matrix. This small deviation gives rise to the parallel Moiré in Fig. 4.40a. From the distance D of the lines and with the above equation, the difference of the interplanar spacings d_1 ($= d(200)$ of η') and d_2 ($= d(200)$ of aluminum) can be determined very accurately: It is 0.0047 $\pm$ 0.0005 nm [20]. This method is of great help in the determination of lattice parameters of unknown structures. Within the *circled area* in Fig. 4.40a one can see an *edge dislocation*. The extra half-plane, ending in the center of the circle, is easily recognizable. According to the schematic of Fig. 4.39 and the above equation, the Moiré effect "images" this dislocation, together with the (200) lattice planes, with about 50 × magnification. This value has to be multiplied with the normal magnification of the TEM micrograph.

Another feature in Fig. 4.40 is of interest. The cubic lattices of matrix and precipitate are three dimensional. In exactly parallel position one

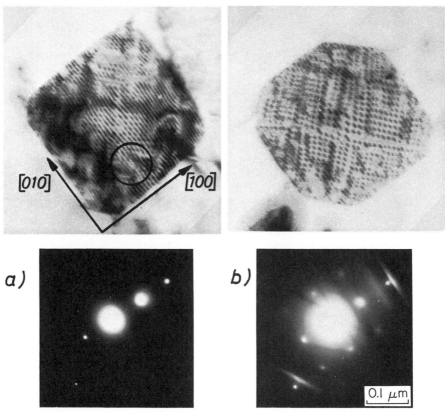

Fig. 4.40. Moiré lines in platelet-shaped precipitates of η'-phase in an Al–0.2% Au alloy [20]. Heat treatment: solution annealed + aged 30 min at 300°C. a) Two-beam case; only the (200) lattice planes are "imaged" by Moiré effect. The circle contains an edge dislocation. b) Multibeam case; a cross grating is formed by the (200) and (020) lattice planes, 110,000×.

would expect, therefore, to see not the pattern of parallel lines drawn in Fig. 4.38a, but rather a corresponding *cross-grating*. The optical analogies for Moiré pattern production were demonstrated in Figs. 4.38 and 4.39. A prerequisite in the EM, however, is that the crystal lattice planes d_1 and d_2 are in reflecting position for the electron beam. In Fig. 4.40b the expected cross-grating is indeed observed for strictly symmetrical orientation with respect to the primary beam. The corresponding diffraction pattern shows the Laue multibeam case. The *line grating* of Fig. 4.40a is generated in analogy to Fig. 4.38, if the sample is slightly tilted from the symmetrical orientation in such a way that only *one* lattice plane system is in reflecting position. As the diffraction pattern shows, Fig. 4.40a is a Bragg case; the Moiré lines are imaged only with the operative **g** vector (200) which is normal to the lines. The details of the precipitation process in AlCu alloys have been clarified by König *et al.* [20a].

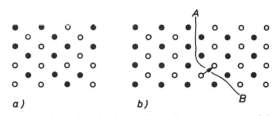

Fig. 4.41. (110) section through a lattice with a) bcc structure and b) CsCl structure. a) Substitutional solid solution crystal with irregular atom distribution in 1:1 ratio. b) Substitutional solid solution crystal with ordered atom distribution. Arrow: displacement vector which would bring the two ordered regions into coincidence. Line *A–B* is the antiphase boundary.

A more accurate treatment of the Moiré effect is possible with the resources of the dynamical contrast theory (see, e.g., Hirsch *et al.* [1]).

4.10. Ordered Structures

In a substitutional solid solution crystal, the different atom species normally are *statistically distributed* among their lattice sites. This case is sketched in Fig. 4.41a for a (110) section through a bcc lattice. It is assumed that the alloy consists of two components with the ratio 1:1. The Fe–50at.% Co alloy is an example; above 730°C it has an unordered bcc structure, designated with α. Below a certain temperature, an *ordered atom distribution* in the lattice constitutes, for some alloys, an energetically more favorable state. This case is also present in the Fe–Co alloy. Below 730°C, the atoms diffuse to the ordered state of Fig. 4.41b. One atom species occupies the cube corners, the other the (equally frequent) cube centers of the formerly bcc structure. Thus is created the *superlattice* α' which, in this case, has the CsCl-type structure. The following alloys have the same type of superlattice: Cu–Zn, Au–Ni, Ni–Al, and Fe–Al. As with x-ray diffraction, so with electron diffraction, the ordering causes additional *superlattice reflections* if the scattering amplitudes of the two atom species are different. The extra reflections are those which because of the extinction rules normally have zero intensity.

The energy of nucleation for ordered regions is small since the energy of the (coherent) interfaces between ordered and unordered regions is also small. Consequently, such nuclei are formed at many places within a grain and then grow together. In the example under discussion, there is a 50% probability that when any two ordered regions (also called *domains*) meet, they will have an identical lattice without a "seam." Such a seam is a so-called "antiphase boundary," as illustrated in Fig. 4.41b. The two different atom species occupy equivalent lattice points in a spe-

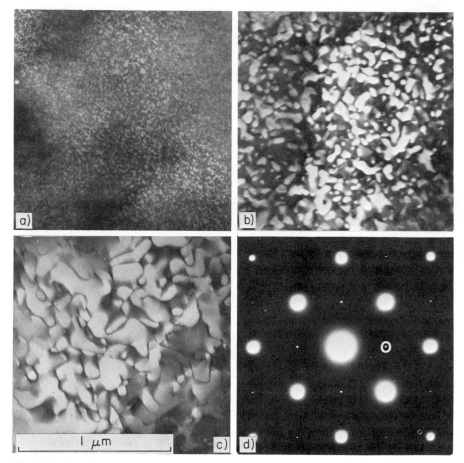

Fig. 4.42. Superlattice domains in an ordered alloy. *Material:* Fe–Co (1:1) with 1.87 at.% V. Heat treatment: homogenized 4 hr at 1100°/H₂O, annealed at 600°C/H₂O. Dark field images in the light of the circled (001) superlattice reflection. a) Annealing time 15 min: Nuclei of the ordered α'-structure, average diameter 10 nm. b) Annealing time 30 min: Ordered regions of α' separated by wide bands of unordered α structure (dark areas). c) Annealing time 4 hr: Antiphase boundaries after coalescence of the ordered regions. All 40,000×. d) [100] diffraction pattern of condition c) (micrographs: D. Krahl, Berlin (Fa. Siemens A. G.)).

cific pattern. Where that pattern changes, the electron wave in the EM undergoes a phase shift. Therefore the antiphase boundary, similar to other two-dimensional boundaries, can produce a fringe contrast unless the boundary lies vertically in the foil. In that case, the antiphase boundary appears as a thin, characteristically bent or winding line.

Figure 4.42 shows TEM pictures of the superlattice domains of a Fe–Co alloy with 1.87 at.% V. Note that the micrographs are *dark field images* taken with a superlattice reflection (circled). Consequently, only the ordered regions are in bright contrast. The series of micrographs shows the growth of domains during an isothermal anneal at 600°C.

4.11. Magnetic Materials

Ferromagnetic materials consist of small regions each having uniform magnetization; these are the so-called Weiss elementary regions. They are separated by Bloch walls where the magnetization vector changes its direction, usually by 90° or 180°. Without special techniques, the Bloch walls are invisible in the light microscope as well as in the electron microscope. In the EM, however, the Weiss regions can be made visible by special procedures.

In Fig. 4.43, the assumption is made that a ferromagnetic foil contains two 180° Bloch walls, and that the magnetization direction is parallel to the foil surface. These directions are indicated by vector tip and end, respectively. During passage of the electron beam through the foil, the electrons are deflected due to the *Lorentz force;* this force acts normal to the primary beam and normal to the magnetization direction. Below each region, the electrons are deflected into opposite directions, as sketched in the figure. Therefore, in a plane *below* the foil (dashed line in the figure), there is an electron deficiency at *A* and an excess of electrons at *B. Consequently, if the EM is focused not on the foil itself, but on the plane (dashed) outside the foil, the Bloch walls are "imaged" as alternately bright and dark lines* (out-of-focus method). The bright and

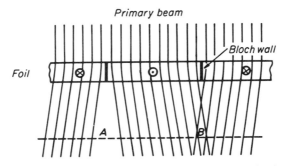

Fig. 4.43. Ray paths through a ferromagnetic foil with the magnetization vectors of the Weiss regions parallel to the foil surface. Because of the Lorentz force the electrons are deflected in such a way that an electron deficit occurs at *A* and an electron surplus at *B*.

dark lines, with reversed sign, are seen also if one focuses on the symmetrical plane *above* the foil. This method is also called "Lorentz microscopy." It has been very useful for the investigation of ferromagnetic materials. An example is shown in the middle micrograph of Fig. 4.44.

The elementary regions can also be made visible in the EM by a second technique, namely, by positioning the objective aperture eccentrically in the direction of the Lorentz force. Because of the beam deflection according to Fig. 4.43, the elementary regions are imaged in alternating lighter and darker contrast. With the help of the ray diagrams, Figs. 1.3 and 1.9, one can easily visualize this effect. A particular advantage of this method is the fact that all object details are in focus, as seen in the bottom micrograph of Fig. 4.44.

4.12. Methods for Improving the Accuracy of Orientation Determination and for Obtaining Unambiguity

a) Improving the accuracy

If Kikuchi lines are present, the crystal orientation, according to Section 3.11, can be determined with an accuracy of $\pm 1°$ or better. Without Kikuchi lines, the accuracy can be improved somewhat by the so-called *center-of-gravity method* as follows: For an exactly symmetrical case ($\mathbf{g} \perp \mathbf{k}_0$, $s = -\theta g$), the diffraction pattern, according to Fig. 4.10, is symmetrical around the primary beam I_0. The more distant spots are weaker because the Ewald sphere intersects only the ends of the streaks. If the crystal lattice is tilted by the angle ϵ from the symmetrical position, the "center of gravity" S of the diffraction pattern (see Fig. 4.45; moves away from I_0. S or S^*, respectively, is that point where the zone axis (ZA) penetrates the image screen or the Ewald sphere, respectively. In reciprocal space $\epsilon = \overline{O^*S^*}/k_0$, and in real space $\epsilon = \overline{I_0S}/L$ (where L is the camera length) and $\overline{O^*S^*} = \overline{I_0S}/\lambda L$ (compare Sections 3.3 and 4.5.1).

The circular area with diffraction spots of the Laue zone in question, and thus the center of gravity S, can be recognized more accurately, the closer together the spots are (i.e., the larger the interplanar spacing d) and the shorter the streaks (i.e., the thicker the crystal). If ϵ is large, the intersection with the Ewald sphere in Fig. 4.45a produces a *ring-shaped area of reflections* whose center is S. With even larger ϵ, the diffraction pattern degenerates further and only a band-shaped portion through I_0 remains of the ring. In that case, the method is no longer applicable; one can only state that the real orientation deviates by a large amount from that determined from the diffraction spots.

The length of the streaks is proportional to $1/t$. It is independent of the indexing and depends only on the crystal thickness t. This fact has the following consequence: During tilting away from the reflecting position, the diffraction spots persist longer, the smaller their

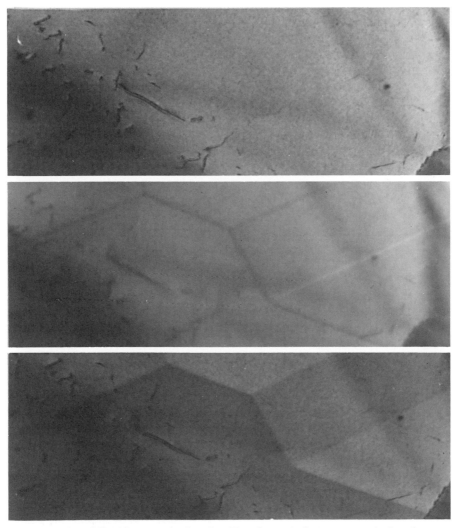

Fig. 4.44. Iron foil with magnetic domains. Top: Image in focus—magnetic domains are invisible. Middle: Overfocused objective lens—domain walls appear as alternating bright and dark lines. Bottom: Excentric position of contrast aperture—domains appear in alternating lighter and darker contrast. 10,000× (micrographs: W. Pitsch, M.P.I. für Eisenforschung, Düsseldorf).

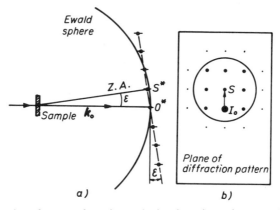

Fig. 4.45. Illustration of center-of-gravity method: a) in reciprocal space and b) in real space.

distance g from the origin I_0, i.e., under otherwise equal conditions, the orientation determination, in principle, is more accurate, the larger g or the higher the indices of the measured reflections. For details see the investigations of Laird et al. [21].

A rather accurate *orientation determination according to Ryder and Pitsch* [22] is possible if a diffraction pattern contains three reflections g_1, g_2, and g_3 from different Laue zones, Fig. 4.46a. A prerequisite is that the three reflections have approximately equal brightness and that they do not form angles near zero or 180°. In that case, the exact orientation **n** is

$$\mathbf{n} = g_1^{\,2}\,(\mathbf{g}_2 \times \mathbf{g}_3) + g_2^{\,2}\,(\mathbf{g}_3 \times \mathbf{g}_1) + g_3^{\,2}\,(\mathbf{g}_1 \times \mathbf{g}_2).$$

n in this equation is a kind of weighted average of the three different vector products resulting from the three reflections, each product representing the crystal normal.

An example is the diffraction pattern, Fig. 3.17a, which has been discussed in detail in Section 3.11. The reflections 511, 151, and 311 fulfill the above prerequisites. From the equation one obtains **n** = [57, 48, 220]. The orientation determined more accurately from the Kikuchi lines in Section 3.11 was [25, 17, 95]. The difference between the two orientations is 2.1°.

b) Unambiguity

The following method is applicable if the diffraction pattern contains *reflections from two Laue zones*. To start with, therefore, those rules should be recalled which govern the correct indexing of patterns con-

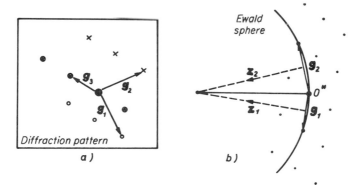

Fig. 4.46. Utilization of the reflections from two Laue zones. a) to increase the accuracy of orientation determinations (after Ryder and Pitsch [22]), b) to obtain unambiguity (after Ryder and Pitsch [23]).

taining several Laue zones (as in Fig. 3.17):

(i) Reflections which belong to two Laue zones simultaneously have the same indices for both zones (as, e.g., $\overline{3}11$ in Fig. 3.17).

(ii) The indices of neighboring reflections have to be neighbors also in the reciprocal lattice, both with respect to numbers as well as signs.

(iii) The zone axes should not enclose a large angle, i.e., the angle should not be more than 10°–20°.

The *180° ambiguity* described in Chapter 3.10 can be eliminated, according to Ryder and Pitsch [23], in all those cases where two Laue zones are present. To accomplish this one makes use of the curvature of the Ewald sphere. The drawing in Fig. 4.46b contains the reciprocal lattice points g_1 and g_2 belonging to two Laue zones with the zone axes z_1 and z_2.[10] Because of the curvature of the Ewald sphere, both angles (g_1,z_2) and (g_2,z_1), are slightly *smaller* than 90°, i.e., the *scalar products* of the vectors are *positive*. According to Chapter 3.10 the 180° ambiguity corresponds to a sign reversal of all indices (hkl), while the signs of the normal $[uvw]$ are retained. The incorrect one of the two indexing possibilities can thus be eliminated with the above criterion, because the scalar products of the incorrect indices are smaller than zero.

Following this rule the indexing example of Fig. 3.17 which contains two Laue zones is unambiguous.

[10] The reflections of two Laue zones drawn in Fig. 4.46b are the same as those in Fig. 3.17.

If reflections of only one zone are present the only alternative is to tilt the sample in the EM until reflections of a neighboring zone appear. Either by using the method described or by three-dimensional visualization of the reciprocal lattice one can then decide which of the two indexing possibilities is the correct one. This procedure is laborious; also it cannot be used as an afterthought. Rather the experimenter has to recognize during observation on the image screen that the diffraction pattern cannot be made unambiguous by any other means. The same method can be used to eliminate the *coincidence ambiguity* unless the latter can be solved by the presence of reflections from a neighboring Laue zone, which is usually the case. (Also see Refs. [17] and [19] to Chapter 3.)

Besides, Kikuchi lines or the center-of-gravity method cannot achieve unambiguity either, unless one of the prerequisites mentioned here or in Chapter 3.10 are fulfilled.

The comprehensive review of the problems and possibilities for exact and unambiguous orientation determination of Ryder and Pitsch [24] is recommended.

The examples in this and the preceding chapters show how the relatively simple model of the Ewald sphere and the reciprocal lattice can be used to derive a relatively large number of statements, methods, and criteria.

4.13. Concluding Remarks

Obviously this short introduction has been very limited in scope. The dynamical theory has only been touched upon. Even concerning the kinematical contrast theory, it has not been possible to discuss in detail its application to all kinds of lattice defects. Instead, it was deemed more instructive to treat in depth the application of the kinematical contrast theory to the case of an *ideal crystal* and to *real crystals with dislocations* (since the latter can be taken as examples for the other types of lattice defects). The conditions in ideal crystals play a role in *all* applications, and dislocations are the most important lattice defects. For example, it appeared more useful to treat the derivation of thickness fringes and bend contours and the thickness determinations based on them (because these effects occur in *all* crystalline samples), rather than, e.g., the determination of the stacking fault energy with the node method (which is of interest only for this *specific* problem).

The goal of this volume as an introduction will be reached if the few but thoroughly discussed examples have enabled the reader to gain easy and quick access to other problems treated in the literature. Besides, in spite of the motto "seeing is believing," the reader should now have acquired the critical eye to recognize when and where electron microscopical phenomena have to be interpreted with special care.

References

1. P. B. Hirsch, A. Howie, R. B. Nicholson, D. W. Pashley, and M. J. Whelan, "Electron Microscopy of Thin Crystals." Butterworths, London, 1971.
2. G. Thomas and M. Goringe, "Transmission Electron Microscopy of Materials." Wiley, New York, 1979.
3. H. Alexander, Z. Metall. **51**, 202 (1960).
4. M. v. Heimendahl, W. Bell, and G. Thomas, J. Appl. Phys. **35**, 3614 (1964).
4a. G. Thomas, in "High-Temperature High-Resolution Metallography." Gordon and Breach, New York, 1967.
5. M. v. Heimendahl, and G. Thomas, Trans. AIME **230**, 1520 (1964).
6. R. Siems, P. Delavignette, and S. Amelinckx, Phys. Status Solidi **2.1**, 421 (1962).
7. P. B. Hirsch, A. Howie, and M. J. Whelan, Phil. Trans. R. Soc. A **252**, 499 (1960).
8. L. Reimer, "Elektronenmikroskopische Untersuchungs- und Präparationsmethoden," 2nd ed., Springer, Berlin and New York, 1967.
9. R. Gevers, Phil. Mag. **8**, 769 (1963).
10. A. Howie, and M. J. Whelan, Proc. R. Soc. London A **267**, 206 (1962).
11. J. A. Bailey and P. B. Hirsch, Phil. Mag. **5**, 485 (1960).
12. C. S. Smith and L. Guttman, Trans. AIME **197**, 81 (1953).
13. R. K. Ham, Phil. Mag. **6**, 1183 (1961).
14. R. K. Ham and N. G. Sharpe, Phil. Mag. **6**, 1193 (1961).
15. U. Eßmann, Phys. Status Solidi **3**, 932 (1963), Acta Metall. **12**, 1468 (1964).
16. F. Pfeifer and I. Pfeiffer, Z. Metall. **55**, 398 (1964).
16a. I. Pfeiffer, Z Metall. **57**, 635 (1966).
17. M. v. Heimendahl, Aluminium **45**, 347 (1969).
18. M. v. Heimendahl and G. Wassermann, Z. Metall. **54**, 385 (1963).
19. M. F. Ashby and L. M. Brown, Phil. Mag. **8**, 1083 (1963), Phil. Mag. **8**, 1649 (1963).
20. M. v. Heimendahl, Acta Metall. **15**, 1441 (1967).
20a. H. König and M. v. Heimendahl, Z. Metall. **70**, 419 (1979).
21. C. Laird, E. Eichen, and W. R. Bitler, J. Appl. Phys. **37**, 2225 (1966).
22. P. L. Ryder and W. Pitsch, Phil. Mag. **18**, 807 (1968).
23. P. L. Ryder and W. Pitsch, Phil. Mag. **15**, 437 (1967).
24. P. L. Ryder, and W. Pitsch, in "Methodensammlung der Elektronenmikroskopie" (G. Schimmel and W. Vogell, eds.), Nr. 4.3.1. Wissenschaftliche Verlagsgesellschaft mbH., Stuttgart, 1971.
25. R. Bullough, D. M. Maher, and R. C Perrin, Phys. Status Solidi (b) **43**, 689 (1971).
26. P. Skalicky and A. Papp, Phil. Mag. **25**, 177 (1972).
27. H. P. Karnthaler, P. M. Hazzledine, and M. S. Spring, Acta Metall. **20**, 459 (1972).
28. D. J. H. Cockayne, I. L. F. Ray, and M. J. Whelan, Phil. Mag. **20**, 1265 (1969).
29. R. de Ridder and S. Amelinckx, Phys. Status Solidi (b) **43**, 541 (1971).
30. F. Schuh and M. v. Heimendahl, Z. Metall. **65**, 346 (1974).
31. K. Schneider and M. v. Heimendahl, J. Mater. Sci. **5**, 455 (1970).
32. E. E. Affeldt and M. v. Heimendahl, Z. Metall. **68**, 595 (1977).

Author Index

Affeldt, E. E., 217
Alexander, H., 92, 143, 149, 160, 174, 217
Amelinkx, S., 217
Andrews, K. W., 45, 139
Ardenne, M. von, 1
Ashby, M. F., 200, 201, 217

Bach, H., 71, 92
Bailey, J. A., 217
Barber, D. J., 71, 92
Beaman, D. R., 46
Bell, W., 139, 217
Bitler, W. R., 139, 217
Blum, W., 92
Boersch, H., 1, 8
Borries, B. von, 1
Bowkett, K. M., 137, 139
Brammar, I. S., 45
Brown, L. M., 200, 201, 217
Bruche, E., 1
Bullough, R., 217
Busch, H., 1

Chandler, J. A., 43, 45
Cliff, C., 43, 46
Cockayne, D. J. H., 217
Cohen, J. B., 95, 139
Cohen, M., 62, 92
Cosslett, V. E., 45
Cullity, B. D., 139

de Broglie, 4
Delavignette, P., 217
de Ridder, R., 217
Dewey, M. A. P., 45
Dick, E., 45
Doig, P., 45
DuBose, C. K. H., 60, 92
Dudek, H. J., 37
Duff, W. R., 62, 92

Duggan, B. J., 139
Dyson, D. J., 45, 139

Eichen, E., 108, 139, 217
Essmann, U., 217

Fischione, E. A., 60, 92
Fleischmann, W., 93
Flewitt, P. E. J., 45
Fuchs, A., 191
Fuerstenau, D. W., 93

Gevers, R., 217
Gillespie, P., 71, 92
Gleiter, H., 92
Glenn, R. C., 62, 92
Glocker, R., 139
Goodhew, P. J., 45
Goringe, M., 44, 92, 217
Gradel, G., 91, 93
Griem, W., 116, 139
Groves, G. W., 70, 92
Grubb, D. T., 69, 70, 92
Grundy, P. J., 45
Guttmann, L., 217

Ham, R. K., 217
Hazzledine, P. M., 217
Heidenreich, R. D., 1, 44
Heimendahl, M. von, 42, 43, 45, 61, 91, 92,
    93, 121, 125, 126, 128, 129, 130, 136, 138,
    139, 152, 154, 166, 192, 204, 217
Hendriks, A., 46
Heywood, J. A., 139
Hirsch, P. B., 1, 4, 17, 44, 48, 52, 92, 139,
    163, 170, 173, 175, 184, 186, 196, 209, 217
Hornbogen, E., 44
Howie, A., 44, 92, 139, 181, 217
Hubred, G. L., 93

Isasi, J. A., 46

Jesser, W. A., 139
Johannson, H., 1
Johari, O., 139
Jones, G. A., 45

Kay, D. H., 45
Keech, G. H., 139
Keller, A., 92
Keown, S. R., 45, 139
Knoll, M., 1
König, H., 208, 217
Karnthaler, H. P., 217
Krahl, D., 210
Krause, F., 1
Krauth, A., 68, 71, 92

Laird, C., 139, 159, 214, 217
Lenusky, J. L., 115, 139
Lewis, M. H., 92
Lonsdale, D., 45
Loretto, M. H., 45
Lorimer, G. W., 43, 46

Machu, W., 92
Maher, D. M., 217
Mahl, H., 1
Makin, M. J., 31, 45
Maurer, K. L., 92
Maussner, G., 45
Mayer, H., 73, 92
Murr, L. E., 44

Nicholson, R. B., 44, 92, 139, 217
Nolder, R. L., 110, 111, 139

Olsen, G. H., 139
Oppolzer, H., 42, 43, 45

Papp, A., 217
Pashley, D. W., 44, 92, 139, 217
Pearson, W. B., 139
Peavler, R. J., 115, 139
Perrin, R. C., 217
Petermann, J., 92
Pfefferkorn, G., 45
Pfeifer, F., 189, 217
Pfeiffer, I., 189, 193, 217

Phillips, V. A., 45, 67, 68, 69, 70, 92
Pitsch, W., 213, 214, 215, 216, 217
Ploc, R. A., 139
Poppa, H., 81
Pumphrey, P. H., 137, 139
Puppel, D., 63, 78, 83, 205

Rack, H. J., 62, 92
Ray, I. L. F., 217
Reichel, R., 189, 205
Reimer, L., 19, 44, 45, 67, 92, 139, 179, 217
Reppich, B., 78, 92
Roser, W. R., 110, 139
Ruska, E., 1
Russ, J. C., 46
Ryder, P. L., 214, 215, 216, 217

Sagel, K., 139
Samudra, A. V., 130, 139
Schäffer, H., 92
Scheucher, E., 62, 63, 92
Schimmel, G., 45
Schmidt, H. D., 83
Schneider, K., 45. 217
Schoone, R. D., 92
Schrader, A., 88, 89, 92
Schüller, K.-H., 84, 85, 86, 87, 92
Schuh, F., 217
Schumann, G., 92
Schwab, P., 116, 139
Schwartzkopf, K., 106, 137, 139
Segall, R. L., 139
Shalicky, P., 217
Sharpe, N. G., 217
Sieberer, K. H., 24, 124
Siems, R., 164, 217
Sinclair, R., 45
Singer, R., 92
Smallman, R. E., 45
Smith, C. S., 217
Soloski, L. F., 46
Spring, M. S., 217
Steigerwald, K., 37
Steinbrecher, H., 73
Stiegler, J. O., 60, 92
Swann, P. R., 2

Tegart, W. J. McG., 52, 92
Theler, J. J., 48, 92

Thomas, G., 4, 44, 45, 48, 92, 93, 106, 110, 111, 139, 150, 155, 206, 207, 217
Thon, F., 45

Urban, K., 30, 45

Vainshtein, B. K., 45
Vierling, G., 81
Vingsbo, O., 125, 139
Vogell, W., 45
von Ardenne, M., 1

von Borries, B., 1
von Heimendahl, M. see Heimendahl, M. von

Warlimont, H., 31, 45
Washburn, J., 44
Wassermann, G., 121, 139, 217
Wegmann, L., 45
Whelan, M. J., 44, 92, 139, 217
Willig, R., 45, 73, 139
Wolff, U. E., 71, 93

Subject Index

Abbé theory, 2, 15, 17, 23
aberration
  chromatic, 6, 10, 12, 30
  spherical, 6, 10, 12, 26
absorption of electrons, 16, 30, 151
acid saw, 48
age hardening, 43, 203
ambiguity
  coincidence, 129ff, 216
  180°, 127ff, 215
amorphous specimens, 15ff, 122
amplitude–phase diagram, 142, 148ff, 160,
  174ff, 178, 191, 193
annular detector, 40
anode of EM, 7ff
antiphase boundary, 209f
aperture
  objective, 6ff, 17ff, 141, 212
  selector, 7f, 26f
aperture angle, 3, 6, 19
artifacts, 76, 84
astigmatism, 6, 10, 33
ASTM card file, 99, 139
atomic number, 17, 19, 30, 36, 42
atomic scattering amplitude, 142, 145, 154,
  161

backscattered electrons, 36, 40
bend contours, 163f, 165ff
Bloch walls, 211
Bollmann technique, 51, 56f, 61
boundaries
  antiphase, 209f
  grain, 25, 86, 141, 149, 151, 163ff, 185,
  202
  low (small)-angle, 188, 191
  tilt, 188
  twin, 149, 151, 163, 165f
  twist, 188

Bragg case, position, orientation, 135f,
  143, 146, 149f, 152f, 155, 157, 161f,
  168, 178, 208
Bragg diffraction, 7, 20ff, 40, 95ff, 122
Bragg equation, 20f, 97f
brick-shaped crystal, 154, 157
bright-field image, 26ff, 37, 125
brittle fracture, 82
bubble formation, 51
Burgers vector determination, 178, 180,
  183ff

calibration standard, 99ff, 113
  of stereo specimen stage, 137
  of tilting stage, 137
camera constant, 98f, 104, 112f
camera length, 97f, 153
carbides, 86, 88
carbon film, 74, 76ff, 80f, 90
cascade generator, 30
cast structure, 47
cathode (of EM), 7ff, 42
cathodes (for polishing), 56f
cell structure, 188, 190
center-of-gravity method, 212, 216
ceramic, oxide, 70, 84, 86
chemical thinning, 48, 71ff
chromatic aberration, 6, 10, 12, 30
cleaving (preparation), 63, 73
Cockroft–Walton generator, 30
coherency, coherent, 165f, 197, 200, 207,
  209
coherency strain, 200, 203
coincidence ambiguity, 129ff, 216
collodion, 75f, 89
column approximation, 172
composite materials, 89
condenser lens, 7f, 33, 35
contamination (of sample), 30, 42, 53, 123

223

contours, 167ff, 178, 180
  bend, 163f, 165ff
  extinction, 25, 164, 180f, 185
contrast, 13, 15ff, 19, 21, 25, 29, 31, 80,
    89, 122, 141, 177, 180, 187, 198
  absorption, 74
  computer simulated, 183
  diffraction, 25, 33, 200
  double, 182
  enhancement, 80
  fringe or stripe, 25, 149, 160, 163ff, 192,
    200, 210
  interface, 200, 202
  mass, 19, 122, 202, 204
  material, 33
  maximum, 177
  orientation, 25, 33, 200, 202
  reversal, 204
  scattering absorption, 19, 122, 200, 202
  stacking fault, 192
  strain, 199f
  structure factor, 19, 74
  topographical, 33
  twin, 192
cooling stage, 30
coordinates
  contravariant, 114
  covariant, 114
corrosion layers, 68
creep, 43, 61, 191
cross slip, 188
crystalline samples, 19ff, 80, 95, 122
crystals,
  ideal, 151ff, 164, 177
  real, 171
current–voltage characteristic curve, 49
cutting, 47, 65
  acid jet, 48

dark-field image, 26ff, 37, 124f, 165, 167,
    194
Debye–Scherrer pattern, 25, 40, 95f, 99ff
deformation, 192, 194
  cold, 47, 188
  texture, 100
  warm (elevated temperature), 191
depth of field, 36
deviation parameter, 151ff, 163, 169f, 171f,
    178f
dielectric coefficient, 85f

diffraction
  angle, 20, 96f
  Bragg, 7, 20ff, 40, 95ff, 122
  constant, 98f
  error, 12
  Fraunhofer, 2f, 23, 26
  pattern, 7, 25f, 95ff
  selected area (SAD), 26, 95, 101
  without lenses, 100f
disk specimens, 59, 187
dislocation, 30, 118, 123, 141, 173ff, 182ff,
    206
  configuration, 188
  density, 185ff
  dipole, 185f
  edge, 178ff, 185, 188, 206f
  forrest, 63
  half, 183, 188
  loop, 31, 182, 203
  mobile, 184
  network, 188, 191
  pair, 182
  partial, 185, 191, 193
  perfect, 183ff
  prismatic, 184, 203
  screw, 173ff, 182f, 188
  tangles, 63f
dispersion hardening, 165
displacement vector, 171, 173, 177, 191
domains
  magnetic, 213
  ordered, 209ff
ductile fracture, 38
dust, 89
dynamical theory, 141ff, 157, 161f, 169,
    177, 180, 182, 204, 209

edge dislocation, 178ff, 185, 188, 206f
elastic scattering, 16, 30, 131, 143
electrochemical deposition, 73
electrochemical thinning, 48
electrolytes for polishing, 52ff
electrolytic deposition, 62
electrolytic polishing or thinning, 47ff, 198
electron
  diffraction, 95ff
  energy, 4, 19, 30
  energy loss, 40, 131
electron microscope
  structure of, 6ff

emission, 32ff
reflection, 32
electrons, backscattered, 36, 40
  secondary, 32, 35f, 40
etch pits, 78f
etching, 49f, 80
evaporation (vacuum), 73ff, 76f, 80
  carbon film, 74, 76
  at normal incidence, 77
  oblique, 77
  of SiO, 82
Ewald sphere, 145ff, 151ff, 157ff, 171, 180,
    196, 212, 215f
extinction, 143
  contours (fringes), 25, 164, 180f, 185
  distance, 151, 161ff, 169, 204
  laws, 95, 105, 209
extraction replica, 86ff

field emission cathode, 7, 42
field ion microscope, 15
float off, 75
focal distance, 6, 9
Formvar, 75
forrest dislocations, 63
fracture
  brittle, 82
  ductile, 38
  intercrystalline, 82
  step, 82
  surfaces, 38, 74, 83
  transcrystalline, 82
Frank partial dislocations, 185
Fraunhofer diffraction, 2f, 23, 26
Frenkel defects, 31

g·b criterion, 177f
geometric optics, 6, 17
GP zones, 160, 197ff
grain boundary, 25, 86, 141, 149, 151,
    163ff, 185, 202
grain size, 86, 95f
grids, specimen support, 51f, 60
grinding, 47f

habit(us), 118, 123, 125, 197, 202
habit planes or directions, 118, 122f, 194,
    196
hammering (preparation), 63

heating of specimen (inadvertent), 17, 30, 123
hexagonal layer, 194
high-temperature examination, 33
high voltage electron microscopy, 29ff, 187
Huygens–Fresnel principle, 141

ideal crystals, 151ff, 164, 177
illumination, oblique, 23, 26
image forces (from surfaces), 30, 187
image formation, 15ff
inclusions, 53, 68
incoherency, incoherent, 166, 197, 202f
indexing,
  consistency of, 112, 114
  of diffraction patterns, 105ff, 113ff
  of directions and planes, 118ff
inelastic scattering, 16, 30, 131
intercrystalline fracture, 82
interface, 200, 209
  contrast, 200
interference fringes, 25, 65f, 164f, 180
intermediate lens, 7f, 25
interstitial type defects, 183, 201
ion etching, 33
ion thinning, 70f
irradiation damage, 31, 73

jet polishing, 58ff, 72

Kikuchi lines, 130ff, 152ff, 159, 169, 180,
    212, 214, 216
kinematical theory, 141ff, 149, 157, 160ff,
    169, 173, 177
Kossel lines, 138

LaB₆ cathode, 7, 42
lacquer, 50, 56, 60, 72, 75
lattice
  defects, 30, 141, 144, 180
  factor, 146, 154
  noncubic, 108, 112, 115, 118
  real, 102f, 108, 144, 194
  reciprocal, 102ff, 113f, 146f, 150,
    153, 157ff, 170, 194ff, 215f
Laue case, 135f, 153f, 155, 158, 178, 180,
    208
Laue pattern, 96
Laue zone, 113, 127, 132, 158f, 212, 214f
lens, 4f
  condenser, 7f, 33, 35
  defects, 6, 10ff

intermediate, 7f, 25
objective, 8f, 17, 22, 25, 27, 31
projector, projective, 7ff, 25
line intercept method, 187
Lorentz force, 5, 211f
Lorentz microscopy, 212

magnetic
   domain, 213
   materials, 211ff
   rotation, 10ff, 29, 116f, 138, 196
magnification, 5ff, 9, 11, 36
mass density, 19
metastable phases, 202f, 207
microanalysis, 41
microtome, 65
Miller indices, 20f, 95, 98f, 110
misfit, 197, 200
Moiré pattern, 65f, 205ff
   parallel and rotation, 206f
mold replica, 74, 80f
Mowital, 75f, 87, 100
multibeam case, 136, 178, 182, 188, 200,
   208

network, 188, 191
nitrides, 86
noncubic lattices, 108, 112, 115, 118

object cartridges, special, 10
objective aperture, 6ff, 17ff, 141, 212
objective lens, 8f, 17, 22, 25, 27, 31
oblique illumination, 23, 26
ordered structures, 182, 209ff
orientation determination, 115, 136f, 159
   accuracy of, 130f, 136f, 159, 212ff
   ambiguity of, 129f, 137, 214ff
orientation relationship, 197, 202
orientation unambiguity or uniqueness,
   127f, 132, 214ff
Orowan mechanism, 203
Ostwald ripening, 43
oxide
   ceramics, 70, 84, 86
   films (to be examined), 68
   layer (contamination), 53
   replica, 74, 79f

parallelepiped, 154ff
particle size, 158
penetration of radiation, 96

perchloric acid, 52, 54f
phase difference, 19f, 144f, 147, 150, 174f
phase shift, 172, 174f, 191, 200, 210
plastic replica, 74ff, 122
platelets, 197ff, 207f
polishing, electrolytic, 47ff, 198
polycrystal diffraction pattern, 95, 99ff
polygonization, 189, 190
pores (voids), examination of, 68, 71
powder metallurgy, 89, 165
powders, preparation of, 89f
precipitates, 24, 28, 47, 53, 113, 117ff,
   121ff, 141, 160, 196f, 200, 207
prethinning, 47f, 61
projection effect, 118, 122f
projector, projective lens, 7ff, 25

ray path in the EM, 8, 18, 22, 27
real crystals, 171
recovery, 188
recrystallization, 47, 63, 165, 190, 194
reflecting sphere, 146f
reinforcement of evaporated films, 82
rel points, 107f, 151f, 157ff, 180
replica,
   carbon, 74, 76f, 122
   evaporated film, 76
   extraction, 86ff
   mold, 74, 80f
   oxide, 74, 79f
   plastic, 74ff, 122
   reinforcement of, 76
   single stage, 74f
   techniques, 74ff
   Technovit, 80f
   Triafol, 80, 84f
   two-stage, 74, 80
resolution, 2f, 6, 10ff, 23, 30, 33, 36ff, 198,
   203
   highest, 23
   line and point-to-point, 13f
Riecke–Ruska lens, 37
ring patterns, 95f, 99ff
rinse, final, 53
R_n ratios, method of, 108ff
roughening of background, 198

SAD, 26, 95, 101
sampling, 84
samples, amorphous, 15ff, 122

saw, 48
scanning electron microscope, 33
scanning transmission EM, 37ff
scattering, 15ff, 30, 96, 141f
  absorption, 202
  amplitude, 142, 144ff, 152, 154, 156, 162,
    174, 191, 209
  angle, 142f
  elastic, 16, 30, 131, 143
  inelastic, 16, 30, 131
Schrödinger equation, 143, 162
screw dislocation, 173ff, 182, 188
secondary electrons, 32, 35f, 40
secondary (subsidiary), intensity maxima,
    157, 168f
selector aperture, 7f, 26f
semicoherency, 197
shadowing of evaporated films, 77, 80
Shockley partial dislocations, 185, 191
single crystal pattern, 95ff
sintering, 89
SiO films, 82
slip traces, 123f, 203
spark erosion, 48, 58
spherical aberration, 6, 10, 12, 26
spot pattern, 95ff
stable phases, 202
stacking fault, 63, 118, 141, 149, 163f,
    183ff, 190ff, 200
stacking fault energy, 188
standard stereographic projection, 119f,
    123f, 136f
standardization, 99ff
stereo specimen stage, calibration of, 137
stereographic projection, 118ff, 127, 129,
    136, 138
stigmator, 10, 33
streaks, 157f, 160, 171, 194, 196, 198f, 202,
    212
structure factor, 96, 161
subgrains, substructure, 188, 191
superlattice, 209ff
suppliers, 58, 61, 80, 91f
support disks (for specimens), 57, 89
support film, 89f
support grids, 51f, 60
surface EM, emission, 32ff
surface EM, reflection, 32
surface layers, 68
suspension medium, 90
symmetrical case, 154f, 158, 212

Technovit, 80ff
texture, 96, 100
thickness
  determination, 120ff, 159, 163, 169ff,
    187
  fringes, 25, 150, 163f, 203f
  of sample, 19
thinning
  acid jet, 48, 71
  aimed, 60f
  chemical, 48, 71
  electrolytic, 49ff
  ion, 70f
Thomas–Fermi–Dirac model, 143
tilt boundary, 188
tilting stage, calibration of, 137
trace analysis, 118ff, 120, 123ff
transcrystalline fracture, 82
Triafol, 76, 80, 82ff
twins, twin boundaries, 141, 149, 151, 163,
    165f
twins
  annealing, 165f
  deformation, 63f, 194, 196
  macro and micro, 194
  recrystallization, 165f, 190, 192, 194
twin structures, 192, 194
twist boundaries, 188
two-beam case, 136, 152, 155, 161, 165,
    169, 178, 182, 200f, 208

ultramicrotome, 65
ultrasonic treatment, 90
unambiguity (or uniqueness) of orientation
    determination, 127ff, 133, 212, 214ff
unit cell, 103

vacancy type defect, 183, 185, 201
vacuum evaporation, 73ff, 76f, 80
vector addition, 105, 108, 112, 114
viscous layer, 51
voids (pores), examination of, 68, 71

washing solutions, 56
wave length, 3f, 30, 96
weak-beam method, 183
wedge fringes, 149, 160, 163ff, 192, 204
Wehnelt cylinder, 8f
Weiss regions, 211
window method, 49ff, 61
wire preparation, 60ff

x-ray
  analysis, 37, 40f
  diffraction, comparison with, 7, 95ff
  mapping, 37, 40
  production in the EM, 16, 31

ZAF correction, 42, 44
zone, zone axis, 97, 115f, 120, 127, 136f,
  177, 215
zones, GP, 160, 197ff